Vegetable Processing

VEGETABLE PROCESSING

Editors

DAVID ARTHEY
Deputy Director

and

COLIN DENNIS
Director-General

Campden Food and Drink Research Association
Chipping Campden, Gloucester

Blackie
Glasgow and London

Published in the USA by
VCH Publishers
New York

Blackie and Son Ltd
Bishopbriggs, Glasgow G64 2NZ
and
7 Leicester Place, London WC2H 7BP

Published in the USA by
VCH Publishers, Inc.
220 East 23rd Street, Suite 909
New York, NY 10010-4606

10 9 8 7 6 5 4 3 2 1

First published 1991

British Library Cataloguing in Publication Data

Vegetable processing
1. Vegetables. Processing
I. Arthey, D. (David) II. Dennis, C. (Colin)
664.8

ISBN 0-216-92907-5

Library of Congress Cataloging-in-Publication Data

Vegetable processing / editors, D. Arthey and C. Dennis.
p. cm.
Includes bibliographical references and index.
ISBN 1-56081-081-5
1. Vegetables—Processing. I. Arthey, D. (David) II. Dennis, C.
TP443. V44 1990
664′.805—dc20 90-1177
CIP

Preface

Wherever in the world temperate vegetable crops are grown, there will be an increasing need to ensure not only their efficient harvesting and rapid handling, but also their preservation – short or long term – so that they reach their destined markets in as near perfect condition as possible. All fresh vegetables begin to perish as soon as they are severed from the growing plant and a multiplicity of technologies has been developed to capture their quality, freshness and nutritional value and to present them in an acceptable form to the consumer. Some of these technologies were exploited in ancient times; others have been developed within the last two hundred years, and some are comparatively recent innovations. Today it is possible to present to the consumer a vegetable product of such high quality, it is barely distinguishable from its condition when it was harvested on the other side of the world.

All principal methods of vegetable processing presently known will remain important, but the technologies employed to exploit them will become increasingly sophisticated in the constant search for improved quality and consistency of production. This book addresses the important and fast moving issues of processing technologies as they apply to vegetable preservation today. It is an up-to-date account of just how much the different techniques have developed over recent years to bring vegetables to the consumer not only in different forms and styles but also with a high degree of safety. The use of microprocessors to control factory operations is a major example of the use of advanced technology which ensures that complicated processes are carried out fully and correctly to guarantee product safety and quality. Also important is the positive move of the vegetable and other food processing industries to adopt new quality assurance concepts which monitor all the key points of the production line. Such quality assurance systems often demand a change of attitude at top management level and affect the philosophy of all key personnel in the factory.

As the world's population expands there is an increasing demand for food, and as that population becomes more affluent so the demand for quality intensifies. The number of different processing techniques now available means that the same vegetable can appeal in many ways to different ethnic groups and a greater percentage of the population can be satisfied. The use of processing techniques also means that hungry populations can be fed with foods that have long shelf lives and yet retain both their value in appeal and their nutritional quality.

As editors, we have between us over forty years of experience in food science and technology, both in the growing and processing of vegetable crops and in their post-harvest physiology. Our contacts with the contributing authors have been built up over those years. They all have wide research and industrial experience and are recognised as experts in their fields. We consider ourselves fortunate to have secured their co-operation and we have drawn on them to provide what we believe is a major contribution to our understanding of vegetable processing. To our knowledge, such a unique collection of in-depth studies of the different aspects of vegetable processing does not exist elsewhere. The chapters comprehensively up-date existing technologies and deal with more recent impacts, such as aseptic packaging, the technology of chilling and the increasingly important area of plant, equipment and cleaning, in detail. We are grateful to all the authors for applying themselves so diligently to give of the best of their knowledge and skills.

David Arthey
Colin Dennis

Contributors

Keith Anderson	Brooke Bond Foods Ltd, Croydon, UK
John Bettison	CMB Packaging Technology plc, Wantage, UK
Chris Brimelow	Westreco Inc., New Milford, Connecticut, USA
Helen Churchill	Campden Food and Drink Research Association, Chipping Campden, Gloucestershire, UK
David Conning	The British Nutrition Foundation, London, UK
Albert Hersom	Consultant in Food Technology, UK
Christopher Knight	Campden Food and Drink Research Association, Chipping Campden, Gloucestershire, UK
David Lyon	Campden Food and Drink Research Association, Chipping Campden, Gloucestershire, UK
David Reid	Department of Food Science and Technology, University of California, Davis, California, 95616, USA
Peter Rutledge	Commonwealth Scientific and Industrial Research Organisation, CSIRO Division of Food Processing, North Ryde, Australia
David Timperley	Campden Food and Drink Research Association, Chipping Campden, Gloucestershire, UK
Dharam Vadehra	Westreco Inc., New Milford, Connecticut, USA

Contents

1 Introduction **1**

David Conning

1.1 Nutritional value 1
1.2 Vegetable processing 3
  1.2.1 Edibility 3
  1.2.2 Reduction of toxins 4
  1.2.3 Preservation 5
1.3 Effects of processing 8
  1.3.1 Nutritional 8
  1.3.2 Socio-economic 8
  1.3.3 Aesthetic 10
1.4 Future of vegetable processing 10
References 11

2 Crop production, harvesting and storage **12**

Christopher Knight

2.1 Introduction 12
2.2 The processing market 12
2.3 Consistent quality and supply 14
2.4 Raw material quality 15
2.5 Factors affecting quality 16
  2.5.1 Choice of cultivar 16
  2.5.2 Crop maturity 17
  2.5.3 Crop husbandry 21
  2.5.4 Harvesting 30
  2.5.5 Storage 32
2.6 Aspects of production of some major vegetable crops 34
  2.6.1 Peas 34
  2.6.2 Carrots 36
  2.6.3 Beans 37
  2.6.4 Brassicas 38
  2.6.5 Potatoes 39
Further reading 41

3 Preparation procedures **42**

Peter Rutledge

3.1 Introduction 42
3.2 Receipt of vegetables into the factory 43
3.3 Conveying 43
  3.3.1 Flumes 44
  3.3.2 Pumped systems 44
  3.3.3 Mechanical conveyors 44

3.4 Dry cleaning 45
3.5 Washing 46
3.6 Inspection 48
3.7 Trimming 51
3.8 Peeling 51
3.8.1 Mechanical methods 52
3.8.2 Lye peeling 53
3.8.3 Hot water 54
3.8.4 Steam peeling 55
3.8.5 Skin removal 56
3.9 Slicing and dicing 57
3.10 Blanching 59
3.10.1 Water blanching 61
3.10.2 Steam blanching 62
3.10.3 Vacuum steam blanching 64
3.10.4 In-can blanching 65
3.10.5 Microwave blanching 65
3.10.6 Hot-gas blanching 65
3.10.7 Texture of blanched vegetables 65
3.10.8 Loss of nutrients during blanching 66
References 67

4 Thermal processing 69

Albert Hersom

4.1 Principles of heat processing 69
4.2 Classification by pH 70
4.3 Heat penetration 72
4.4 Containers 74
4.4.1 Cans 74
4.4.2 Glass containers 75
4.4.3 Flexible pouches 75
4.4.4 Trays 76
4.4.5 Thermally intolerant containers 76
4.5 Outline of canning operations 76
4.5.1 Preparation of the food 77
4.5.2 Filling 78
4.5.3 Exhausting the container 78
4.5.4 Sealing 79
4.5.5 Thermal processing (sterilisation) 79
4.6 Aseptic processing and packaging 84
4.6.1 Sterilisation of the product 84
4.7 Nutrient retention during thermal processing 97
4.8 Storage of canned products 98
References 100

5 Freezing 102

David Reid

5.1 Introduction 102
5.2 The freezing process 103
5.2.1 Heat exchange materials 104
5.2.2 Characteristics of the vegetable 105
5.3 Blanching 107

5.4 Industrial freezing 108
5.4.1 Batch freezers 111
5.4.2 Semi-continuous freezers 114
5.4.3 Continuous freezers 115
5.5 Product characteristics 117
5.5.1 Peas 118
5.5.2 Green beans 118
5.5.3 Carrots 119
5.5.4 Broccoli 119
5.5.5 Cauliflower 119
5.5.6 Brussels sprouts 120
5.5.7 Spinach 120
5.5.8 Sweet corn 120
5.5.9 Potatoes 120
5.6 Storage 121
References 122

6 Chilling **123**

Chris Brimelow and Dharam Vadehra

6.1 Introduction 123
6.2 Historical and commercial aspects of chilled vegetables 124
6.3 Process treatments for minimally processed vegetables 125
6.3.1 Storage temperature 126
6.3.2 pH modification 127
6.3.3 Water activity 128
6.3.4 Surface coating systems 128
6.3.5 Surface dipping systems 129
6.3.6 Addition of competitive microflora 130
6.3.7 Heat treatment 130
6.3.8 Speed of chilling 131
6.3.9 Sanitary processing and packaging 132
6.3.10 Controlled atmosphere and modified atmosphere packaging 133
6.3.11 Irradiation processing 134
6.4 Manufacturing methods 137
6.4.1 Preparation methods 137
6.4.2 Heat-processing methods 137
6.4.3 Chilling methods 139
6.4.4 Sanitary packaging methods 140
6.4.5 Manufacturing methods for catering outlets 142
6.5 Distribution and retailing 143
6.6 Nutritional and quality aspects 145
6.7 Safety aspects 147
6.8 Conclusions 150
References 150

7 Other preservation methods **154**

Keith Anderson

7.1 Introduction 154
7.2 Dehydration 154
7.2.1 Water activity 155
7.2.2 Field drying and sun drying 156
7.2.3 Air drying 156
7.2.4 Spray drying 157

7.2.5 Vacuum drying 158
7.2.6 Freeze drying 159
7.2.7 Dehydrofreezing 160
7.2.8 Intermediate moisture food and osmotic drying 160
7.3 Dried vegetables 160
7.3.1 Preparation 160
7.3.2 Legumes 161
7.3.3 Brassicas 162
7.3.4 Root vegetables and tubers 162
7.3.5 Others 162
7.3.6 Storage and packaging 163
7.4 Salt 163
7.5 Salting and brining 164
7.5.1 Salting 164
7.5.2 Fermentation brining 164
7.5.3 Cauliflower brining 165
7.5.4 Cucumbers and gherkins 166
7.5.5 Onions 167
7.5.6 Red cabbage 168
7.5.7 Walnuts 168
7.5.8 Other vegetables 168
7.5.9 Sauerkraut and the sauerkraut fermentation 168
7.5.10 Olives 169
7.5.11 Fermentation inhibition in brining 170
7.6 Freshening or debrining 170
7.6.1 Continuous methods 170
7.6.2 Multibatchwise methods 171
7.6.3 Single-batchwise methods 171
7.7 Pickling 171
7.7.1 The preservative effect of acetic acid 172
7.7.2 The influence of other acids 173
7.7.3 Raw materials 173
7.8 Pasteurisation of acid products 176
7.9 Preservatives 178
7.9.1 Sulphur dioxide 179
7.9.2 Benzoic acid and the alkyl-4-hydrobenzoates (parahydroxybenzoates) 180
7.9.3 Sorbic acid 180
7.10 Syruping 180
7.11 Hurdle and combination techniques 180
7.12 Colours 181
7.13 Enzymes 181
7.14 Packaging 182
7.15 Spoilage of acid products 182
7.15.1 Multiplication of yeasts, moulds or bacteria 182
7.15.2 Deterioration through oxidative reactions 183
7.15.3 Deterioration due to action of trace metals 183
7.15.4 Deterioration due to interaction of component substances 184
7.15.5 The presence of foreign matter or contaminants 184
7.16 Shelf-life 184
References and further reading 185

8 Packaging **186**

John Bettison

8.1 Introduction 186
8.2 Objectives of packaging and market considerations 187
8.2.1 Packaging requirements 187

8.2.2 Economic benefits 188
8.2.3 Marketing benefits 188
8.2.4 Long-term benefits 189
8.3 Containers for heat-preserved vegetables 189
8.3.1 Metal cans 190
8.3.2 Glass containers 197
8.3.3 The retort pouch 208
8.3.4 Rigid/semi-rigid plastics containers 212
8.3.5 Aluminium foil trays 217
8.3.6 Collapsible tubes 218
8.3.7 Aseptic packaging 218
8.4 Packaging of frozen vegetables 224
8.5 Packaging of dehydrated vegetables 227
8.6 Packaging of fresh/chilled vegetables 228
Further reading 229

9 Quality **230**

David Lyon and Helen Churchill

9.1 Introduction 230
9.1.1 Definition of quality 230
9.1.2 Lifestyle and trends 231
9.1.3 Consumers 232
9.1.4 Perceived quality 232
9.2 Factors influencing product quality 233
9.2.1 Selection of raw materials 233
9.2.2 Post-harvest treatment and handling 234
9.2.3 Processing and presentation 234
9.2.4 Packaging 237
9.2.5 Storage and distribution 238
9.2.6 Consumer handling and use 240
9.2.7 Nutritional aspects – processed versus fresh 240
9.3 Measurement of product quality 241
9.3.1 Instrumental 241
9.3.2 Near infrared spectroscopy 245
9.3.3 Immunoassay 246
9.3.4 Sensory evaluation 246
9.4 Product specifications 248
9.4.1 Deriving specifications 248
9.4.2 Types of specification 251
References 252

10 Plant, equipment and cleaning **254**

David Timperley

10.1 Introduction 254
10.2 The location 254
10.3 The site 254
10.4 The building 255
10.5 Equipment 260
10.6 Installation 266
10.7 Maintenance 266
10.8 Cleaning 268
References 274

Index **275**

1 Introduction

DAVID CONNING

1.1 Nutritional value

Plants are the source directly or indirectly of all food, essentially because they are capable of using solar energy and of various chemical syntheses which cannot be accomplished by animals. Prime among these are photosynthesis, the use of solar energy to synthesise carbohydrates, and nitrogen fixation, the basis of protein synthesis.

The main consumption of plants has concentrated on cereals, legumes and tubers, which have become dietary staples. The term 'vegetables' includes a wide range of plants and plant structures. In considering their nutritional value, they are most easily classified as follows:

(i) *Leaves* – leafy vegetables, such as cabbage, sprouts and lettuce.
Stems, such as asparagus.
Specialised leaves, such as leeks, onions and celery.

(ii) *Legumes* – seeds within pods, such as peas, beans, and lentils. Seeds, when dried, are termed *pulses*.

(iii) *Tubers*–such as potatoes and yams.

(iv) *Roots*–such as carrots, turnips and beetroot.

(v) *Miscellaneous*–cucumber, tomato and mushrooms.

Most vegetables contain at least 60% water and usually more than 80%. Their value as sources of macronutrients (protein, fat, carbohydrate) is limited but there are important exceptions. Legumes, for example, are rich in starch and protein; tubers are rich in starch. Apart from legumes and tubers, the main nutritional value of vegetables is their content of micronutrients (vitamins and minerals), and poorly digestable complex carbohydrates (dietary fibre) which have very little nutritional value but are important for bowel function. The nutritional content of examples of each class (Tables 1.1, 1.2 and 1.3) illustrate that individual items cannot provide all the requisite nutrients, but by different combinations – that is, by eating a variety of vegetables – the complete nutritional range can be consumed (1). This is of particular importance with respect to two nutrients, protein and essential fatty acids. Proteins are polymeric molecules constructed from a range of amino acids. There are about 20 amino acids that are important to mammalian biology. During digestion, proteins are split into their component amino acids and, after absorption, the

Table 1.1 Nutritional value of vegetables: Macronutrients.

Type	Energy (kcal)	Nutrient (%) Protein	Fat	Carbohydrate	Water	Fibre
Cabbage	26	3.3	—	3.3	90	3
Peas	67	6	—	11	79	5.2
Potatoes	87	2	—	21	76	2
Carrots	23	—	—	5.4	90	3
Tomatoes	14	1	—	3	93	1.5

Table 1.2 Nutritional value of vegetables: Micronutrients (per 100 g).

Type	Carotene (μg)	Thiamin (mg)	Riboflavin (mg)	Nicotinic acid (mg)	Vitamin C (mg)
Cabbage	300+	0.06	0.05	0.3	60+
Peas	300+	0.32	0.15	2.5	25+
Potatoes	—	0.11	0.04	1.2	10+
Carrots	12 000	0.06	0.05	0.6	6+
Tomatoes	600+	0.06	0.04	0.7	20+

Table 1.3 Nutritional value of vegetables: Minerals (μg per 100 g).

Type	Sodium	Calcium	Magnesium	Iron	Zinc
Cabbage	23	75	20	0.9	0.3
Peas	1	15	30	1.9	0.7
Potatoes	7	8	24	0.5	0.3
Carrots	95	48	12	0.6	0.4
Tomatoes	3	13	11	0.4	0.2

requisite proteins are reconstructed. There are ten amino acids that cannot be synthesised in mammalian tissues and which must be present in the diet. These are therefore termed *essential amino acids*. In practice, there are also four amino acids of prime importance nutritionally because their content in commonly consumed foods may be limited. These are lysine, methionine, cystine and tryptophan. Their content in common foods items (Table 1.4) is such that deficient intake is unlikely if a varied diet is consumed.

There are two essential fatty acids, linoleic acid (18:2 n–6) and linolenic acid (18:3 n–3). These are used in the synthesis of messenger substances that influence arterial tone and blood coagulability. They are currently under investigation for their roles in coronary heart disease. Some vegetables are very good sources of these fatty acids (Table 1.5). Vitamin E is included in this

Table 1.4 Relative amino-acid content of some foods (mg per g N).

Foods	Lysine	Methionine	Cystine	Tryptophan
Cabbage	190	60	70	60
Peas	470	60	70	60
Potatoes	340	100	80	90
Carrots	350	120	—	90
Tomatoes	180	40	40	50
Wheatflour	150	100	160	70
Beef	570	170	80	80
Fish	610	180	70	70

Table 1.5 Essential fatty acids in foods (mg/100 g food).

Food	Linoleic acid	Linolenic acid	Vitamin E
Cabbage	—	—	0.2
Peas	30	8.5	—
Potatoes	50	10	0.1
Carrots	—	—	0.5
Tomatoes	—	—	1.2
Beef (lean)	90	60	0.15

table because it is an important dietary antioxidant. The essential fatty acids, being polyunsaturated, are vulnerable to oxidation with the possible production of peroxy free radicals which may cause tissue damage. The most potent dietary sources of essential fatty acids and vitamin E are the vegetable oils derived from seeds and nuts.

1.2 Vegetable processing

Processing is undertaken for three reasons:

(i) To render the produce edible
(ii) To reduce the content of plant toxins
(iii) To preserve the produce – that is, to prevent spoilage due to autolysis or microbiological attack.

1.2.1 *Edibility*

The application of heat – cooking – is the commonest form of processing. It fulfils all the objectives of processing but, for vegetables, is used primarily to render the produce edible. Heat application can be in an aqueous medium (boiling, pressure cooking), in oil (frying) or by microwave. In each case the effects are the gelatinisation of starch granules, the hydrolysis of hemicellulose, and the protopectins are rendered soluble by depolymerisation. In addition, the cell walls become permeable, the cells lose water and the leaching of nutrients, lectins and other components occurs. These effects are accentuated at alkaline pH.

In frying and microwaving, hydrolysis is effected by the water content of the vegetable so that the final produce has a better texture and losses are minimised.

1.2.2 *Reduction of toxins*

Plants produce a variety of toxins, probably as a defence against insect predators. In addition, they contain compounds which are part of the normal plant chemistry but which are potentially toxic to man. Examples are the phenolic and cyanogenetic glycosides. Some potentially toxic compounds that occur in vegetables but which are destroyed or leached by heat processing are given below.

(i) *Proteins*: Phytohaemagglutinins that cause clumping and destruction of red cells occur predominantly in beans. Some of these compounds are also toxic to the cells of the intestinal mucosa. They are normally destroyed by soaking the beans overnight followed by thorough cooking.
(ii) *Lathyrogens*: Aminonitriles and diaminobutyuric or propionic acids that cause spastic paralysis as a result of prolonged consumption of inadequately processed chickpeas in excessive amounts. This arises usually as the result of the unavailability of other food.
(iii) *Glucosinolates*: Sulphur compounds that occur in brassicas (cabbage and turnips etc) which can cause goitre (enlargement of the thyroid gland).
(iv) *Enzyme inhibitors*: Mainly trypsin inhibitors that occur in soya and lima beans.
(v) *Cyanogens*: Classically in cassava root but, of more relevance to western diets, also found in lima beans.
(vi) *Phenolics.*
(vii) *Oxalates.*

A group of enzyme inhibitors, the anticholinesterases – which include glycoalkaloids α-solanine and α-chaconine – are resistant to heat processing. They occur in the potato and are mainly associated with green skins and shoots but on storage may migrate throughout the flesh. They are detected by their bitter taste. Usually peeled potatoes cause no problem, residual concentrations being too small (2).

The levels of toxins in edible plants have been reduced by plant breeding programmes, but with a consequent increase in susceptibility to pests. The use of wild varieties to improve 'domesticated' species has caused problems with enhanced toxin production.

1.2.3 *Preservation*

Two predominant characteristics that have determined the successful socio-economic development of the human species have been the domestication of animal and plant species and the ability to preserve food. The latter preceded the former by many millennia for there is evidence that primitive man learned that heated food remained edible for longer than raw food and that meat and grain could be preserved by drying. Later the ability to preserve large stocks of grain between harvests represented a significant advance, which quickly extended to fruit. The empirical discovery that salt, saltpetre and acid media enhanced this ability was a later development, but it was not until the work of Pasteur had laid the foundation for a scientific understanding of food spoilage that the beginnings of modern preservation technology were established.

Food spoilage is due to two distinct processes:

(1) *Autolysis* – the digestion of food by enzymes present within the tissues which are released when the cell membranes lose their integrity as a consequence of death.
(2) *Microbial attack* – the invasion by bacteria and fungi.

The nature of these processes dictates the actions required to protect the food (Table 1.6) but essentially preservation techniques are directed at four fundamental biological properties:

(i) Enzymes are essential biological catalysts without which living processes are impossible. Enzymes are specialised proteins.
(ii) All active organisms require water.
(iii) The hydrogen-ion concentration must be within a fairly narrow range of pH. For most organisms this is pH 6 to 8. Some organisms can survive acidity less than pH 4.
(iv) Nucleic acids are essential to the metabolic processes of active life. If they are disrupted, cell death results.

The main preservation processes, therefore, depend on:

- Reduced enzyme activity, either by heat denaturation of protein or by chilling
- Exclusion of water, by osmosis or freezing or drying
- Reduction of pH
- Irradiation (disrupts nucleic acids)
- Atmosphere control

1.2.3.1 *Heat.* When heated above 60 °C, protein undergoes an irreversible change. Proteins are polymeric molecules folded upon each other in specific ways and this folding is essential to their biological function. On heating, folding is disrupted, many cross-links are formed and the protein is rendered less soluble and unable to function. This is the process of 'denaturation'. Enzyme activity is terminated when the enzyme protein is denatured.

Table 1.6 Preservation processes for vegetables.

Type of process	Methodology
Heat	Blanching: Water at 100 °C 2–3 min } plus sodium sulphate, optional Steam 1 2 min } plus sodium sulphate, optional Microwave Cooking: Boiling Pressure cooking Steaming Frying Baking Microwave Canning: Heat and pressure
Water exclusion	Dehydration: Warm air stream Spray drying (soups) Osmosis: Immersion in sodium chloride solution (brine) Freezing: below −18 °C
Enzyme inhibition	Chilling; below 4 °C Blanching Atmosphere control packaging
Acidity	Fermentation: usually combined with brining (pickling) Acetic acid Sodium nitrite Sodium sulphite Sodium metabisulphite Benzoic acid and derivatives
DNA/RNA disruption	Irradiation Chemical: e.g. Tecnazine

The rate of enzyme activity is temperature dependent so that at very low temperatures the rate is markedly reduced. Although microorganisms vary in their ability to function at low temperatures, all activity is markedly slowed at temperatures approaching 0 °C and terminated at −30 °C. For the majority of organisms, activity ceases at −18 °C.

1.2.3.2 pH. Most enzymes function effectively at hydrogen-ion concentrations between pH 6 and 8. It is exceptional for microorganisms to function at a pH less than 4. Such organisms exist however, and they are used extensively in pickling processes around the world. It is probably the case that pickling by fermentation is second only to heat as the commonest form of preservation practised world wide. The organisms used are mainly *Lactobacillus planarum* and *Leuconostoc mesenteriodes*, organisms that generate lactic acid to up to 1.5% concentration with pH values around 3.7. Fruits and vegetables have an advantage in that the pH of cell fluids is slightly acid, which inhibits many bacteria. Microbiological spoilage of vegetables is mainly due to moulds (fungi) but soft rotting bacteria can cause serious losses under certain conditions.

Two organisms which cause food poisoning deserve special mention. *Clostridium botulinum* is a ubiquitous bacterium that is capable of anaerobic metabolism and therefore thrives in the absence of oxygen. It is important because it produces a potent nerve toxin that causes botulism, a disease associated with high mortality due to respiratory failure. Under adverse environmental conditions, this organism can form spores that may remain viable for long periods of time and may revert to reproductive forms when conditions improve. The spores are heat resistant and it is normal practice to subject them to a process equivalent in lethality to a temperature of 121 °C for at least three minutes to kill them. The organism is sensitive to acid conditions and cannot normally grow at a pH less than 4.5. Effective protection against this organism, which can contaminate many types of food, is dependent on the use of sufficient sodium nitrite or sulphur dioxide to generate adequate concentrations of nitrous acid or sulphurous acid respectively, or canning, where high-temperature sterilisation can be achieved (3).

Listeria monocytogenes is an organism present in soil and widely distributed in foods. Contamination of foodstuffs such as vegetables with this bacterium cannot be avoided, but the organism is acid and heat sensitive and is killed by pasteurisation. It is, however, a facultative pathogen and in sufficient numbers can cause disease in people whose defence mechanisms are compromised or undeveloped. The elderly, infants, and patients with immunological deficiency are vulnerable. It is capable of crossing the placenta and infecting the conceptus, usually resulting in abortion. *Listeria* is of particular concern because it can grow at refrigeration temperatures; chilled foods must be thoroughly reheated before consumption. In passing, it may be noted the organism has been found in soft cheese where the acidity is insufficient to kill it. This problem should be circumvented by adequate pateurisation of milk and avoidance of post-process contamination.

1.2.3.3 *Ionising irradiation.* This is cytotoxic by disrupting DNA and RNA (nucleic acid) replication. At dosages not exceeding 10 kGy it is capable of reducing the bacterial and mould burden to negligible levels and inhibiting sprouting. Higher doses are required for sterilisation but carry the risk of inducing organoleptic changes. Food irradiation is permitted in several countries for specified foodstuffs and purposes, and is under consideration for use in the UK.

1.2.3.4 *Packaging and atmosphere control.* Fruit and vegetables continue to live after harvesting and thus generate carbon dioxide and heat, and consume oxygen. By their nature, preservation in the fresh uncooked state is a relatively short-term requirement and can be effected by packaging to control the availability of oxygen and carbon dioxide in the space in which the produce is stored. Reduced oxygen turnover and increased carbon dioxide are required but have to be regulated carefully to avoid anaerobic respiration, which hastens spoilage. Alternative procedures use inert gases, for example, nitrogen, and partial vacuum. For longer periods of storage – that is, more than 2–3 days – reduced temperature is also required for vegetables grown in temperate climes. Those of tropical origin may be susceptible to degeneration if cooled below critical temperatures.

Under development now are plastic films which exhibit differential diffusion gradients for oxygen and carbon dioxide. These hold promise in achieving 'controlled atmosphere' conditions within the narrow limits required.

1.3 Effects of processing

There has been a good deal of public comment about food processing and its consequences on the diet. Such comment has usually concentrated on the nutritional, socio-economic and aesthetic aspects of processing.

1.3.1 *Nutritional*

Any form of processing adversely affects the nutrient content of foods, especially in respect of vitamins. In general, macronutrients may be broken down to their component parts, but this is a process continued by digestion, and losses occur only as a result of leaching. These can be recouped if, for example, stock is used to make sauces and gravy. There are some losses of the amino acid lysine due to thermal degradation, but these losses are not usually of nutritional significance.

The vitamin content of vegetables is very variable depending on season, soil fertility and time and conditions of post-harvest storage. Some vegetables are more susceptible to such losses than others. In general between a third and a half of the vitamin content of raw fresh produce is lost during storage and cooking, unless the produce is deep frozen in the raw state, in which case the losses are ameliorated considerably (Table 1.7). The content may be increased as a result of slight dehydration.

In practical terms, these inevitable losses may be of concern only for vitamin C, for which it is advisable to include some raw produce, mainly fruit or fruit juice, in the diet to ensure adequate intake. Deep-frozen produce, however, retains high levels of vitamin C (4).

1.3.2 *Socio-economic*

Today most developed populations live in urban conurbations and are dependent on an extremely complex food distribution system. It is a tribute to that system that the variety of food available today is greater than it has ever

Table 1.7 Illustrative examples of vitamin C lost during the processing of peas.

Type of process		Retention of vitamin C (dry weight basis) Amount (mg/100)	%
Raw		130	100
Cooked	– 12 minutes – minimum water	116	89
	– after 1 day storage (room temperature)	90	69
Frozen	– three months and cooked (4 minutes)	61	47
Canned	– three months	40	31

been and yet the cost of food continues to decline as a proportion of national expenditure, essentially because increases in price do not keep pace with inflation or earning power.

The supply of fresh vegetables – that is, unprocessed vegetables at around 4–5 days since harvesting – and of deep-frozen vegetables with improved nutritional content has increased in response to the ready availability of chilling facilities (i.e. refrigerators and deep freezers) in the home. In addition, there is now a much expanded range of canned goods with the twin advantages of being ready-cooked and microbiologically safe.

Modern lifestyles demand more convenience in food preparation and consumption, including the ultimate domestic convenience of eating out. Over the last decade or so there has been an overall decline in the amounts of vegetables purchased, but within this there has been an increased demand for frozen vegetables. Given the reduced waste of processed vegetables and the increasing tendency to eat out, the overall consumption of vegetables has probably remained fairly static. There have been substantial changes in the types of vegetables selected, reflecting the much expanded range available. Frozen potato products and frozen peas have increased at the expense of fresh and canned produce, whereas canned dry beans continue inexorably upwards as do potato crisps. Fresh leafy vegetables have declined except for salad varieties which have increased, but there has been increased consumption of root vegetables, onions, leeks and several others. Vegetables have declined as the source of vitamin C, due probably to the substantial increase in consumption of fruit juices.

Without an accurate measurement of food consumed outside the home it is not possible to compare the actual consumption of vegetables over the last decade, but the indications are that the overall consumption remains much the same per head of population. There has been an increase in the range of vegetables eaten and a shift towards processed, especially deep-frozen, products and away from fresh vegetables, although even here there are exceptions (5).

1.3.3 *Aesthetic*

Vegetables are easily overcooked so that ready-prepared meals and canned goods that may be subjected to a high temperature for safety reasons may not meet with approval because of their texture. However, most people seem prepared to trade pristine organoleptic properties for satisfaction, convenience and reasonable taste sensations. Indeed, many canned products – especially beans and soups – have achieved popularity because of their own unique taste characteristics.

On a more pragmatic level, there is no evidence that classic presentation and taste are associated with better nutritional value, though dietetic variety is essential to both and in that sense an educated palate may be an advantage. Most people eat to live and achieve a sense of well-being in other ways. Against this there is the tendency to increase expenditure on food as affluence increases, and there are now many influences on food choice (journals, television, foreign travel, eating out) that will affect food selection. It may be that in many of us there is a gourmet awaiting an opportunity for expression. This could be an important factor influencing future development in food products.

1.4 Future of vegetable processing

Two influences will affect eating habits in the next few years. First, the continued concerns with the effect of diet on susceptibility to disease allied to improved consumer education will result in continued efforts to persuade consumers to moderate their food intake, especially of high-energy foods. Secondly, as a result of the burgeoning popularity of convenience foods and meals, these will increasingly form substantial parts of the diet. Both influences are likely to result in greater consumption of vegetables.

The recognition that the intake of antioxidant vitamins, vitamin C and vitamin E, may have a protective role against cancer will mean that vegetables – a potent source of the relevant vitamins – will be in demand, but this will have to be in a form that protects the vitamin content. It is likely, therefore, that the market for fresh frozen vegetables will outpace other forms of processing in the short term.

Canned products will still command a market but increasingly this will be in the form of combined products that complement a meal, rather than individual items. For example, it is envisaged that the complete vegetable accompaniment of a meat-based meal could be available in a can.

Although the range of vegetables will increase, the main market will still be for conventional items, but in different forms – that is, cooked and presented differently.

As the emphasis will be on providing the consumer with vegetables as near to their natural state as possible, it is not expected that new processes will be needed although an expansion of new forms of packaging can be expected, especially those that allow easy processing, say, by microwave.

Overall, it may be predicted with confidence that vegetables, never highly popular with the consumer, will at last assume their rightful role as an indispensable part of a properly balanced, highly nutritious diet and that there are good opportunities for food producers in the combination of convenience and palatability allied to safety.

References

1. Paul, A.A. and Southgate, D.A.T., McCance and Widdowson's *The Composition of Foods*, HMSO, London, 1978.
2. *Toxicants Occurring Naturally in Foods*, National Academy of Sciences, Washington, USA, 1973.
3. Ranken, M.D. (ed.), *Food Industries Manual*, Leonard Hill, Glasgow and London, 1984.
4. *Food Processing*, British Nutrition Foundation Briefing Paper No. 11, 1987.
5. *House Food Consumption and Expenditure*. Annual Report of the National Food Survey Committee, HMSO, London, 1987.

2 Crop production, harvesting and storage

CHRISTOPHER KNIGHT

2.1 Introduction

Crop production is central to the making of many foods and it is vital to recognise the importance of the raw material quality. A high-quality processed product can only be produced from raw materials that are specifically suitable for the intended market and of high quality because once the crop is harvested that quality at best can only be maintained. The primary producer is, therefore, an integral part of a single food production line. As the first stage in this process, the production of vegetables in the field holds the initial key to the successful marketing of processed vegetable products.

In developed countries, the production of vegetables for processing is highly organised. What was once largely a manual small-scale operation has become an intensive and highly mechanised agri-business in which the areas of production are well defined and developed. Some raw materials may still be grown on small acreages, such as market garden enterprises, but most vegetables for processing are increasingly produced in large-scale and specialised operations more akin to agriculture than horticulture. Such specialised production for the processing market has gone hand in hand with the growth of a large and increasingly sophisticated food-processing industry. As processing technologies have improved and the demand for processed products has increased, the demand for high-quality raw materials has become more important.

2.2 The processing market

Canning and freezing are by far the most common methods of preserving vegetables long term and such products now form part of our everyday life. Dehydration is probably the next most important method of preserving vegetables, and other methods, such as pickling – where the product is preserved with the addition of acetic acid – is a specialised technology whose products occupy a smaller section of the processed vegetable market in some countries.

The canning industry was probably the first to utilise vegetables on a large scale. The modern high-speed canning industry was developed in the early part of this century. The traditional three-piece can, with at first soldered and in more recent years welded side seam, is being replaced with two-piece cans. Apart from these metal cans other materials can be used for the heat preservation of vegetables, and glass is commonly used in mainland Europe. Heat retortable plastic trays and the laminated flexible pouch are modern developments but both of these containers are more likely to be used for recipe dishes than vegetables alone. The basic steps for the heat processing of vegetables are cleaning, peeling or other form of preparation such as slicing, filling, covering with liquor, closing, retorting, cooling and labelling. All these operations require suitable raw materials in order to maximise the quality in the canned product.

The freezing industry has developed more recently than the canning industry. The commercial freezing of vegetables first became acceptable in the 1920s and 1930s, notably in the USA; in the UK it was not until the late 1940s that the commercial freezing of vegetables became established. Since that time the industry has grown steadily, and continues to do so today. The most popular frozen vegetable in the UK is the potato, with green peas second; and the quality of the raw material for freezing is of paramount importance for these and other crops.

Many vegetables can be dehydrated artificially. The simplest process employs the use of forced hot air but a more sophisticated method is to freeze-dry, where the product is first frozen and the water removed by sublimation under a vacuum. The product retains its original size and shape and reconstitutes readily when water is added. Although the dehydrated product does not require a sophisticated package (other than to retain a gas atmosphere) to maintain its quality the drying process is energy intensive. Dehydration remains a minor method of preserving vegetables compared to freezing and canning.

As processing technologies improved and demand for the processed product increased, the demand for high-quality raw material intensified. This has had a marked impact on the vegetable producer or grower. The development of specialised vegetable production for the processing market has gone hand in hand with the growth of an increasingly sophisticated food-processing industry, such that crop production is now an integral part of the food production chain.

In the early days of the food-processing industry the processor, especially the vegetable canner, often had to resort to the use of raw material from the wide range supplied to the fresh market. This was invariably inconsistent both in quality and reliability of supply. As processing technologies improved and demand for products increased it became vital for the processor to obtain a raw material that consistently met the quality required and was of sufficient volume for efficient production.

It was soon realised that if a reasonable quantity of high-quality processed product was to be produced then some control was necessary over the production of the raw material and the processing industry started to work closely with growers. This was most obvious in the UK with the advent of the frozen food industry in the late 1940s although similar developments took place in the canning industry and has since been emulated by most sectors of the processing industry. Contracts were increasingly used to stipulate the quantity and quality of the raw material required. As growers sought to meet these requirements the quality of processed products improved.

Significant resources were committed by the processors to the production of raw materials. Efforts were made to develop the most appropriate methods of production, to select the best varieties and provide a field advisory service. Growers had to work more closely with the industry to meet the exacting quality and volume requirements of processors such that there is now close liaison between processors and growers and between grower cooperatives supplying vegetables. The processing of vegetables is now an important market for growers and they have responded by making the necessary changes in husbandry, harvesting and handling in order to meet the demand for suitable raw materials.

Other sectors of the agricultural industry have also been stimulated to meet the associated demand for more suitable and higher yielding cultivars, better and more efficient harvesters and handling equipment and agrochemicals to ensure the produce is free from blemishes caused by pest and disease attack. Plant breeders have considerable breeding programmes with suitability for processing as a principal aim. Likewise the agricultural machinery industry has been in the forefront of the mechanisation of harvesting. This is such an important aspect that plant breeders have included suitability for machine harvesting as a breeding objective for crops such as green beans and Brussels sprouts.

These and similar developments over a wide range of vegetables grown for processing have led to significant improvements in the growers' ability to produce crops for the processing market.

2.3 Consistent quality and supply

The processing industry requires a consistent supply of fresh high-quality raw material during a predetermined period of production, usually at the height of the appropriate season for each vegetable. This is now achieved by specifying what they want, when they want it and in what quantity. It has since been emulated by other sectors of the vegetable market to one degree or another.

Contracts and specifications are in wide use. Contracts are used to stipulate the quantity of product required, the price and other conditions relating to supply. A specification on the other hand relates to the level of quality of the product and should be referred to in the contract.

Responsible vegetable processors have specifications that define precisely the qualities they are seeking and how the quality can be measured both in the raw material and final product. The specification may define a level of quality well in excess of any legal requirement. One series of specifications that has been developed for the vegetable-freezing industry has been prepared by the Campden Food and Drink Research Association (CFDRA) in the UK. Twenty product specifications have been published and many have gained wide acceptance both within the UK and in international trade.

Each CFDRA specification consists of two parts. The first deals with raw material quality. This section states quality in general terms, details the sampling procedure and gives clear definitions of defects and the levels of tolerance for each and the size grades if appropriate. It may also list acceptable cultivars. The second part defines the quality of the final product for several grades, including the level of quality required to meet the minimum acceptable grade of final product quality.

The advantages of a recognised and well-defined specification are twofold. Firstly, the grower of the raw material knows the quality required of him and that if his delivery of raw material to the processor fails to meet the minimum level of quality it may be rejected. Secondly, the processor knows that if the raw material meets or exceeds the specification when assessed on delivery to the factory he should be able to produce an acceptable final product.

2.4 Raw material quality

Quality is a widely used term. Quality can be good or it can be poor and in terms of raw material it can mean suitability for one market and not for another. Simply, it means fitness for purpose or the degree of excellence. Quality can also be defined in great detail, for example in a specification. In this chapter, quality is defined as the status of the raw vegetable, which is assessed from the relative values of several characteristics that determine the acceptability of the product to the processor.

Raw material quality can be assessed by many parameters but the most common are the number and extent of defects, size, shape, colour, flavour and texture although sometimes others may be considered, including nutritional value and style of presentation – for example, whole cauliflower heads or floretted heads. The individual quality characteristics that may be assessed are numerous and their relative importance will differ with the crop and end use.

Freedom from defects is of paramount importance. This includes freedom from the ravages of insect infestation and microbial contamination. Other symptoms of contamination include off-flavours and taints. Chemical residues such as those arising from the use of pesticides may occasionally be a problem. More specific defects include the presence of foreign matter such as stones, toxic berries and slugs, the presence of extraneous vegetable matter, and blemished off-coloured and damaged produce.

The visual quality of vegetables is well defined and if it is poor then the reasons are invariably known. Other less readily definable aspects of quality are, however, increasingly important. Enhanced eating quality and, particularly, flavour is one example. There is also a greater interest in those aspects of quality concerned with production, which relate to environmental and health issues. The increasing demand for organically grown produce is a notable example. With this type of product the method of production is the paramount quality factor although traditional visual quality aspects remain important.

2.5 Factors affecting quality

Vegetables may be marketed in many different forms and it is essential to know the requirements of the chosen market before the crop is grown. This is because the required raw material quality may differ even for the same crop. The grower must, therefore, pay particular attention to the husbandry of the crop and post-harvest handling of the produce to ensure it is suitable for the chosen market. Similarly, the end user of the produce must give due consideration to the requirements of the grower when specifying the raw material required. Only in this way can high yields of good-quality produce be produced.

The quality of processed vegetables is dependent upon a number of production factors. The importance of these factors differs with crop and the end use. Two crop production factors, however, are of prime importance, and these are choice of cultivar, and maturity or stage of ripeness at harvest. Other factors concerning the husbandry of the crop can also influence quality in certain circumstances. Post-harvest factors can also have a profound influence and quality can only be maintained at best following harvest.

2.5.1 *Choice of cultivar*

Modern vegetables have all been bred from wild species. Some more or less resemble their wild ancestors but the majority are completely different. The brassicas are a notable example of the latter. The Brussels sprout is completely unrecognisable from the wild cabbage plant from which it was derived. Modern cultivars, or varieties as they are otherwise known, are bred for high yield, resistance to disease, suitability for mechanical harvesting, uniform maturity, acceptable quality and many other characteristics in order to meet the requirements of the grower and end user. Growers are aware of the many different cultivars available to them, not only from the seed companies but also research establishments which test and compare the performance in the field (Figure 2.1) and the harvested produce for its suitability for different methods of processing.

The growth and development of vegetable plants can to some extent be modified by the environment including the growing conditions but it is the genetic make-up of the plant which is the principal factor determining the quality and other characteristics of the vegetable product. The choice of the

Figure 2.1 Potato cultivar trial (courtesy of the Potato Marketing Board).

most appropriate cultivar for a particular use is essential. The selection of an inappropriate cultivar makes it difficult or almost impossible to attain the desired raw material quality. The processor, therefore, will benefit from a detailed knowledge of both the agronomic performance and suitability for processing of the cultivars of vegetables that are processed.

A simple but extreme example of the influence of cultivar on quality is shown by broad beans for canning. White-flowered cultivars produce pale green coloured products whereas coloured flowered cultivars produce brown products. Both types if frozen give a pale green colour. Thus the choice of cultivar is critical for the canned product.

2.5.2 *Crop maturity*

Maturity or stage of ripeness at harvest is the other major crop production factor to have a considerable influence on the suitability of raw material for a particular use. Most vegetables are harvested at immature stages of maturity. Few vegetable crops for processing are taken to full maturity, the most notable exception being dry legumes (pulses). This means that the crop is in full growth when it is harvested and yield is increasing. Vegetables toughen as they mature and while yield is increasing the vegetable product is losing tenderness. Thus, there is a conflict between the interests of the grower and those of the processor. The grower wishes to maximise his returns by obtaining as high a yield as possible, but the processor on the other hand needs a high-quality and tender raw material. A compromise, however, is reached where the crop is harvested at a stage of maturity that is acceptable to both parties. Sometimes a sliding scale of payment is arranged for fast maturing crops such as green peas where the grower receives the greatest financial return by harvesting the crop at the desired maturity.

The speed of change or development of maturity varies enormously with the crop and product. For those products which are derived from the reproductive phase, such as the flower or developing seed, this is a period of rapid physiological development and for crops such as immature green beans and peas harvest is judged by some form of maturity index. The exact point of harvest for these products is critical in order to obtain the desired raw material quality. For slower maturing products such as root crops and leafy vegetables, where the vegetative parts of the plant are harvested, the rate of development is much slower and the need to measure maturity is less critical; harvest can be judged by eye.

Crops harvested fully mature are allowed to senesce naturally and the point of harvest is dependent on factors other than just maturity. Dried legumes, for example, are harvested when the seed has dried to or near the level at which the seed can be stored safely. Final drying can be done artificially after harvest if necessary. Nowdays senescence may be encouraged by the use of artificial chemical desiccants which kill the plant without affecting the seed, or tuber in the case of potatoes. This can be a useful crop management tool and potatoes in particular are often treated in this way.

The rate and time at which crops mature is largely dependent on genetic factors and in many crops different cultivars are grown to give a spread of maturity. For example, vining peas are grouped into early, mid-season and late types and different Brussels sprout cultivars are grown for different harvest periods. Similarly, potato cultivars are generally grouped by maturity according to their most useful time of marketing as First Early, Second Early or Maincrop. First Early varieties are lifted for immediate sale, Second Early varieties are also lifted for immediate sale, or when fully mature, may be stored as Maincrops. Maincrop cultivars are lifted fully mature and may be stored for several months. Other factors, such as the time of planting, crop husbandry and the prevailing weather, may have a modifying influence on maturity.

One fast maturing crop that has been studied in great detail from the maturity aspect is the immature green pea (*Pisum sativum*) when it is grown for processing by canning or freezing. Investigations of the composition of the peas have shown that as peas mature, sugars become converted to starch. The optimum maturity for canning is the point of development immediately preceding the beginning of the acceleration of the conversion of sugars to starch (Figure 2.2). This is called the practical canning stage (PCS). This point can be measured chemically by analysis of the alcohol insoluble solids (AIS) of the peas. However, chemical analysis of maturity is impractical in the field and

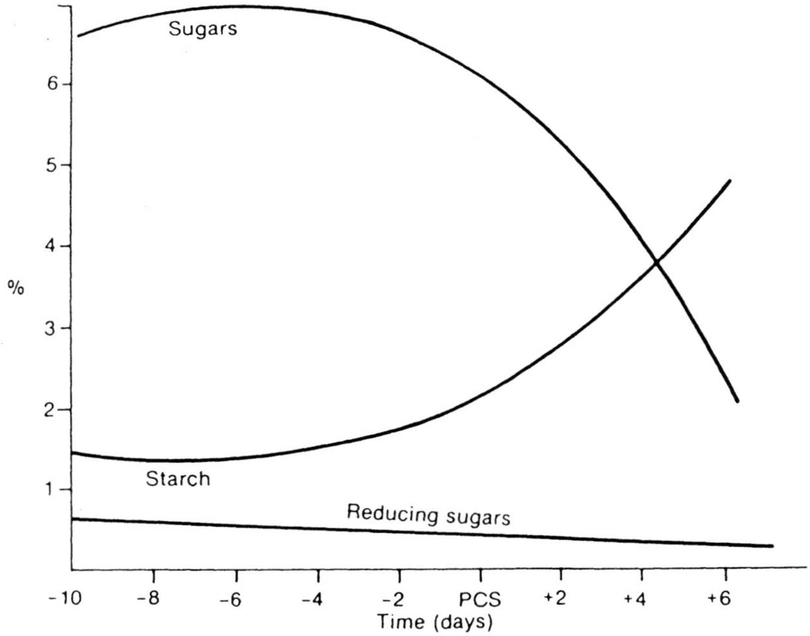

Figure 2.2 Changes in sugars and starches in peas as they mature (Anthistle(1)).

a more rapid physical test has been developed which measures pea texture. The instrument is called a tenderometer (Martin Pea Tenderometer) and is widely used by both growers and processors (Figure 2.3). The PCS corresponds to a tenderometer reading of 120 which approximately corresponds to an AIS content of 13% (Figure 2.4). The tenderometer works by shearing a sample of fresh peas; the force required to shear the sample is indicated on an arbitrary scale. Other instruments have been developed in other parts of the world to measure this important index. The Ottawa Pea Tenderometer, the Tenderometer TD 250 and the FMC Digital Tenderometer are some examples.

As well as measuring the maturity of the pea crop the tenderometer can also be used to predict the optimum maturity of a pea crop so the processor can plan his pea production programme. Harvest date can also be predicted from tenderometer readings. To measure the maturity of the broad bean (*Vicia faba*) however, a slightly different method is employed. Because broad beans are tougher than peas a smaller quantity is used in the tenderometer cell than peas.

The green bean (*Phaseolus vulgaris*) is another relatively fast maturing crop in which the whole seed pod is harvested immature for processing by canning and freezing. The tenderometer is not usually used to measure maturity of green beans and in this instance the length of the developing seed within the

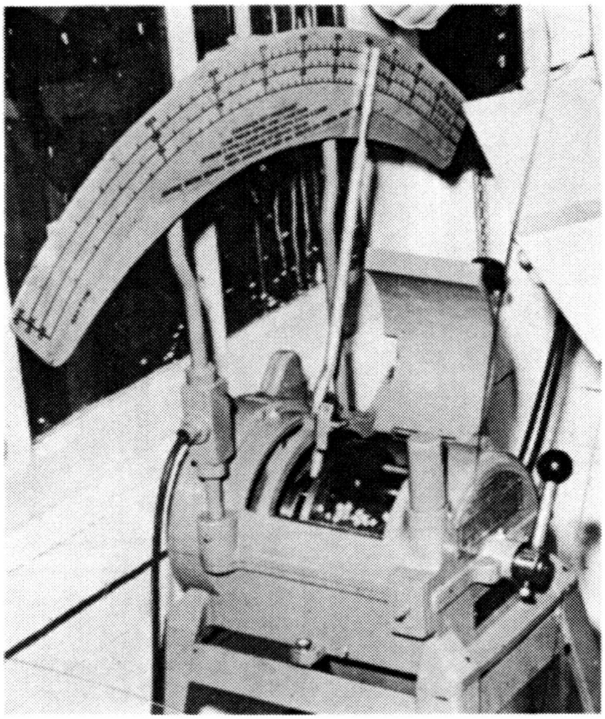

Figure 2.3 Pea tenderometer.

pods is a satisfactory method. Maturity indices have been developed based on seed length. In practice, as the crop is judged to be approaching the appropriate stage of maturity, ten mature plants are taken from different points in the field, the most mature pod is removed from each plant and the largest seed is taken from each pod. The ten seeds are aligned end to end and the length recorded. Because green beans may be long, intermediate or short podded, different indices are required for each type (Table 2.1).

As maturity can be modified by prevailing weather conditions, time of planting and growing conditions, a skilled grower or factory operator should use the information provided by these indices as a guide to maturity and modify it in order to use it to its best advantage.

No similar methods exist for assessing the maturity of other vegetables as they tend to change more slowly as they mature. Crops such as carrots and Brussels sprouts are harvested when judged by eye to be ready. A possible exception is potatoes for canning which are harvested usually when the dry matter is below 20%. In this instance sample digs are taken from the crop and specific gravity measured by a flotation technique using a hygrometer. There is a direct relationship between specific gravity and dry matter; a specific gravity

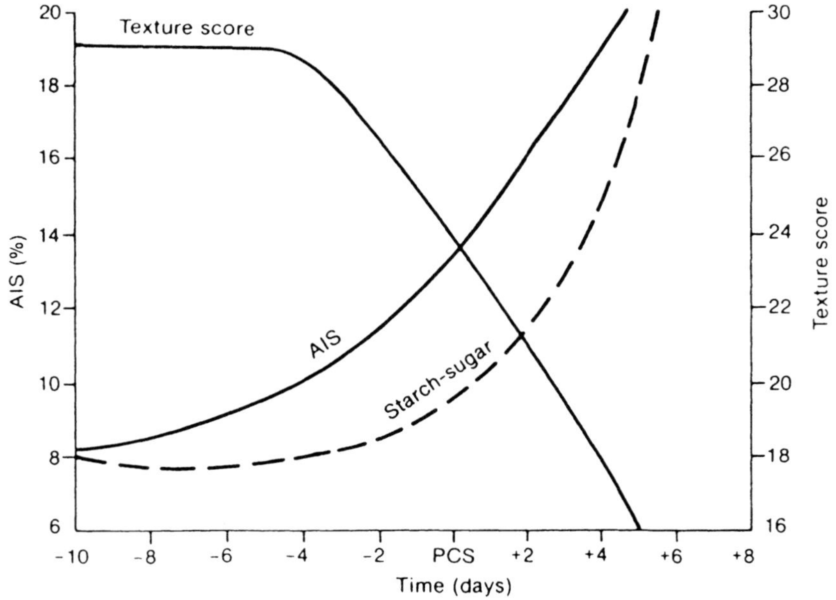

Figure 2.4 Changes in alcohol insoluble solids and texture scores in peas as they mature (Anthistle(1)).

Table 2.1. Optimum maturity indices for French green beans (10 seeds).

Bean type	Canning: Whole pods (mm)	Canning: Cut/sliced pods (mm)	Freezing: Whole pods (mm)	Freezing: Cut/sliced pods (mm)
Long podded	—	120	—	100
Intermediate podded	100	110	80	90
Short podded	100	—	80	—

of 1.080 is equivalent to 20% dry matter. The measurement of specific gravity is preferred because it is for all intents and purposes instant, whereas that for dry matter can take several days to determine.

2.5.3 *Crop husbandry*

In addition to cultivar and maturity, other crop husbandry factors may influence yield and quality of vegetables. Their relative importance will differ with the crop and their influence may be modified by prevailing weather conditions and cultivar. However, these crop husbandry factors should be taken into account when producing a vegetable crop for processing.

2.5.3.1 *Soil type.* Site is invariably an important factor in the production of vegetables and it is essential to grow certain crops in the most suitable areas. Both soil type and the prevailing weather conditions are the two most important contributory factors which decide the suitability of a particular site and different vegetable crops have different requirements. The main production areas for vegetables are often well defined and developed.

Some vegetable crops grow more readily on certain soil types than others producing a higher yield with the harvested produce having a better quality. Crops such as potatoes and carrots are particularly susceptible to soil conditions. Sandy soils tend to be earlier in cropping and they may give a better skin finish to root and tuber crops. New or early potatoes for canning are often grown in lighter soils for this reason. Stoney soils may cause distortion of roots such as carrots and may lead to damage during the harvesting of crops. Potatoes are particularly susceptible to this form of damage. The level of the water table may cause differences in growth patterns of carrots; a low water table will cause the carrot roots to elongate whereas the same variety grown in a soil with a high water table may have shorter roots. Peaty soils are also generally favourable for vegetable production, especially those which require a continuous supply of moisture such as celery. Carrots also do well in a peat soil as root development is unhindered and skin finish is good. For direct sown crops, such as green beans, emergence from germination is better in soils which do not form a hard cap or seal on the surface after rain. Growing seedlings find it difficult to break through hard surface layers.

2.5.3.2 *Plant population.* Plant population or the number of plants in a unit area, may have a direct effect on the size of the plant and the harvested produce. This aspect of production can be controlled to some extent in certain crops. In fact, a number of vegetable crops are manipulated in this way to produce a product of the desired size. In general, the closer the plants are spaced in the field the smaller the size of the harvested product. As with other crop husbandry factors the prevailing weather and growing conditions can have a further modifying influence on the final result.

The production of brassicas is influenced by plant population. Brussels sprouts, for example, are planted at different spacings for different markets (Figure 2.5). Small sprouts are required for the freezing market and the plants are established at a higher density than those destined for the fresh market which will accept a larger sprout. Similarly, the size of cauliflower heads can be controlled by spacing of the plants in the field and very close spacing will produce so called 'mini' cauliflowers.

Some root crops are also influenced by spacing of the plants in the field and this aspect of production is used to produce vegetables suitable for processing. The carrot is a good example, where it has been shown that root size can be controlled by seed rate and hence plant population. In general, the closer the carrot plants the smaller the mature roots will be. Methods have been

Figure 2.5 Uniform plant spacing of transplanted brassicas (courtesy Monsanto plc).

established to calculate the seed rate necessary to produce a high proportion of the crop in the desired root size. Similarly, potato size can also be controlled by the spacing of the seed tubers in the field. Small potatoes are required for canning and the plants are spaced more closely than those required for other markets such as those destined for the French fries and crisping markets which require a larger tuber. Not only are the tubers placed closer together but sprouting is encouraged to produce as many stems as possible which further exaggerates the effect of spacing.

2.5.3.3 *Plant nutrition.* In general, correct plant nutrition and fertiliser treatment are essential to maximise yield and obtain the best quality product. Unless the plant has the necessary major and minor nutrients its development may be restricted and defects may occur. Vegetables are no exception to this rule.

The major nutrients are nitrogen, phosphorus, potassium, magnesium, sulphur and calcium. They are present in soils and are taken up by plants in relatively large quantities. The major elements have a role in building the structure of plants, whereas the minor nutrients or trace elements such as copper, zinc, manganese, boron, molybdenum and iron are important in enzyme systems and contribute to the plant's function rather than its structure. Minor nutrients are essential for plant growth and occur in very small amounts in soils. The three nutrients needed from inorganic fertilisers in greatest quantity are nitrogen, phosphorus and potassium. Calcium is important in determining soil pH and in general a pH of 6.5 for vegetables is recommended.

An insufficient supply of any nutrient can lead to a restriction in crop growth and/or quality. The application of inorganic fertilisers forms an important part of modern crop production for very few soils are able to supply sufficient quantities of all the nutrients necessary for high yields and quality in crops. However, efficient use of fertilisers depends on careful assessment of the total level of the nutrients required by the crop and the amount available from the soil.

All other factors being equal, adequate nutrition ensures acceptable crop growth and quality. Problems can arise, however, if nutrients are in short supply. Nitrogen deficiency, for example, is very common and all vegetables except possibly legumes are susceptible with a yellowing on lower leaves. Manganese deficiency results in the seeds of peas showing 'marsh spot' when split open, which is a serious quality defect. Both deficiencies can be corrected by timely applications of the appropriate nutrients. Conversely, excessive use of fertilisers can be damaging. Some minor nutrients are toxic to plants at high concentrations. Copper, for example, when present in excess replaces other metal ions, particularly iron. In this case root growth is restricted and chlorosis occurs. Excessive use of nitrogen is not only wasteful but can lead to excessive leaching into drainage water with all the consequent environmental problems. For some crops and cultivars this can also lead to higher levels of nitrate in the plant.

Calcium is an essential nutrient, particularly for leafy vegetable crops and a number of common disorders are related to the availability of calcium such as internal browning of Brussels sprouts and blackheart of celery. Similarly, root crops are prone to calcium related disorders such as storage breakdown of potatoes which is often associated with bacterial soft rot. Adequate lime status is essential because it reduces soil acidity – most vegetables will not thrive under acidic conditions – and it increases the availability of certain plant nutrients in the soil. Overliming can cause problems. At high pH values trace element availability is substantially reduced. Boron deficiency, for example, in brassicas is frequently associated with high pH.

2.5.3.4 *Pest, disease and weed control.* Pests, diseases and weeds have important implications for both the grower and user of the harvested produce. They may cause considerable reductions in the potential yield of the growing crop both directly and indirectly such as causing problems at harvest. They may also impair the quality of the produce. Similarly, if the harvested crop is stored they can cause additional damage during storage, further reducing marketable yield and quality. Apart from the obvious effects on growth of the plant, pests and diseases may blemish the produce and contaminate the harvested produce. Both blemishing and contamination are serious quality defects.

Weeds will compete with crop plants for light, water and nutrients, and this may reduce yield or cause the crop plant to grow uncharacteristically. In addition, weeds may cause considerable difficulties at harvest which may indirectly further reduce the recoverable yield. Weeds may cause uneven development of the crop and hence uneven ripening or maturity and choke the harvesting machinery so as to make this operation less efficient. Both these aspects may affect quality of the produce. Weeds, however, have a more direct influence on quality. Extraneous vegetable matter is defined as any harmless vegetable matter, and this includes parts of other plants in the crop, and any weeds present at harvest, therefore, will exacerbate this aspect of quality. Flower buds (for example, thistles) and seed heads (for example, poppies) can be found in vined peas, and, while not causing a toxicity hazard, will downgrade the quality of the raw material and the canned or frozen product will suffer if they are not removed in the factory. Certain weeds produce toxic berries such as nightshade and potatoes and these pose greater problems. In fact potatoes are an important weed in UK pea and green bean crops as they are often included in the same rotation and some potato varieties produce a large number of berries. These berries are readily picked up by mechanical harvesters and processors have to take great care to exclude them from their products.

Similarly, pests will not only reduce yield but may affect quality by causing defects in the produce. Some pests will affect the growth of the plant without directly affecting the produce. Other pests may cause damage to produce before harvest but not be present with the harvested produce – for example, carrot root fly and pea moth in dry harvested peas (Figure 2.6). Yet other pests may be present even in the harvested produce – for example, pea moth in vining peas and caterpillars and aphids in brassicas. If the insect is present in raw material, either dead or alive, this is regarded as a serious quality defect. The presence of such insects is described as foreign matter. If, on the other hand, only the effects of the insect are present then the defect comes under the heading of blemish. While a certain level of blemish may be tolerated the tolerance for foreign matter is invariably nil.

Like weeds and pests, diseases will also reduce yields and affect quality. Some disease-causing organisms will affect the growing plant and reduce yield but do not directly influence the harvested produce, such as pea wilt caused by strains of the fungus *Fusarium oxysporum*, chocolate spot (*Botrytis fabae* and *Botrytis cinerea*) of broad beans, leaf rot of onions (*Botrytis squamosa* and *Botrytis cinerea*) and stalk rot of sweet corn (*Fusarium* spp). Others will directly affect the harvested produce and reduce quality by causing blemishes to the produce. Again, there are many examples on vegetables, notably cavity spot of carrots caused by *Pythium* spp and common scab and powdery scab of

Figure 2.6 Pea moth damage in dry harvested peas.

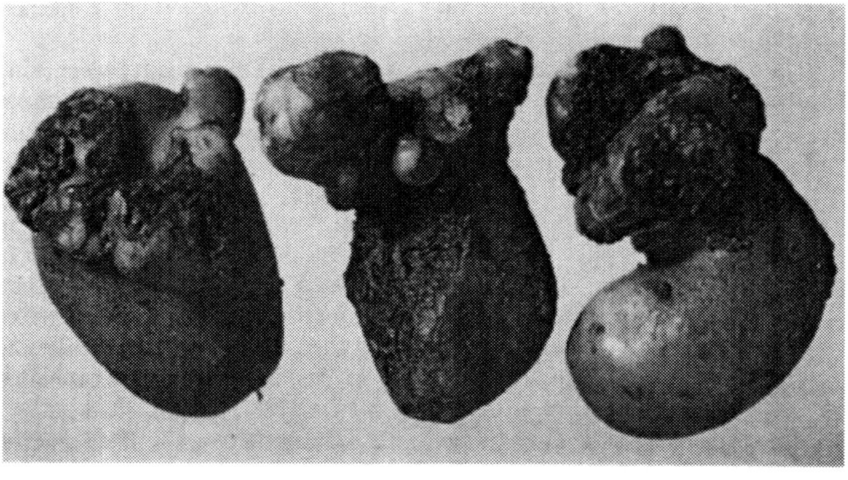

Figure 2.7 Powdery scab of potatoes (courtesy Potato Marketing Board).

potatoes caused by *Streptomyces scabies* and *Spongospora subterranea* (Figure 2.7) respectively, brassica diseases such as dark leaf spot on Brussels sprouts caused by *Alternaria* spp and leaf spot of celery (*Septoria apiicola*). Grey mould (*Botrytis cinerea*) is a very common fungus and will affect many crop plants. It is a serious disease of green beans in particular causing a localised rot on the beans which is classed as a blemish. Infected pods must be removed at the factory and if the level of infection exceeds 5% of pods then the crop is difficult to utilise. Very few disease-causing organisms are obvious contaminants of the harvested produce such that they may be considered foreign matter. One exception which has caused problems for processors in some seasons is the presence of fungal sclerotia in vined peas. These are hard black resting bodies of the fungus *Sclerotinia sclerotiorium* which affects the pea plant. The sclerotia are formed on the surface of the plant, and being of similar size to the peas are readily picked up by the viners along with the peas. These must then be removed at the factory.

Some bacteria and viruses are also important pathogens of vegetable crops affecting both yield and/or quality. Halo blight of French beans caused by the bacterium *Pseudomonas phaseolicola* infects both the foliage and the developing pods. Similarly, many viruses affect the development of the plant and cause a blemish of the vegetable product. Tomato mosaic virus, for example, produces a mosaic pattern on the leaves and a brown/bronze discoloration of the fruit. Turnip mosaic virus symptoms on cabbage include dark necrotic rings and spots on older leaves and in stored white cabbage similar symptoms can also develop in store even though they were not evident when the heads were put in store.

The grower must take all necessary measures to protect his crops from pests, diseases and weeds if he is to maximise his returns and the processor is to receive a high-quality raw material which is blemish free and uncontaminated by extraneous vegetable matter and foreign material. There are three basic elements to any pest, disease and weed control strategy. These are: resistant cultivars, cultural practices and the use of pesticides.

Many vegetable cultivars have been bred for resistance to disease. For some diseases this is a very effective means of control but for many diseases control is not total and has to be augmented by other control measures. For example, resistance to pea wilt is a very important characteristic of cultivars and gives complete control to this important soil-borne disease. This is fortunate as there are no other practical control measures on a field scale. However, the causal fungus (*Fusarium oxysporium* f. sp. *pisi*) occurs in a number of different forms and resistance to the forms present in any particular soil must be present in the cultivar if control of the disease is to be achieved. Downy mildew of peas is caused by another soil-borne fungus *Peronospora pisi* and can also be controlled by cultivar resistance but the control is only partial. Total control can be achieved by growing resistant cultivars and combining this with a pesticide seed treatment to give extra protection to the seedling when it is most vulnerable to infection. Infection can also occur in the growing crop but this is less serious on resistant cultivars.

Resistance to pests in vegetable crops is much less common and, other than rare examples, is not an effective means of control. One notable exception is resistance to the root cyst eelworm (potato cyst nematode) in potatoes. Cultivars of potatoes have been specifically bred to give resistance to this important pest. Resistance to weeds is to all intents and purposes non-existent. Control of weeds, therefore, is entirely dependent on other control measures.

Cultural practices are a useful means of reducing the incidence of pests, diseases and weeds and are an important part of any control strategy. These cultural practices include rotations and different cultivation techniques. Rotations have traditionally been used to maintain soil fertility and keep pests, diseases and weeds at manageable levels. The principle being to keep susceptible crops, which may be hosts to pests and diseases in particular, as far apart in time as is practicable. This helps to ensure that the populations of pest and disease do not build up to damaging levels. Similarly, different weeds are important in different crops and a planned rotation will help reduce or keep their populations in check. Cultivation practices such as ploughing will help reduce the populations of weeds and also bury material that might harbour pests and diseases.

Cultural practices on their own are unlikely to give complete control and must be used in conjunction with other methods. One of these is the use of pesticides. The advent of chemicals to control pests, diseases and weeds had a major impact world wide on the production of vegetables, not least for the processing market. The use of pesticides has enabled the grower to produce maximum yields of defect-free and and high-quality produce and consistently so, such that the processor has an almost assured supply of raw material. Pesticides, therefore, have a significant role in the production of high-quality vegetables.

Today, agrochemicals have been developed to a very sophisticated level. Selective herbicides can be used in vegetable crops which will leave the crop plants unaffected but effectively eradicate all undesirable weeds. Herbicides generally have either a residual effect where they work over a long period inhibiting the germination of weed seeds (Figure 2.8), or a knock-down effect where existing weeds are killed. Similarly, insecticides and fungicides to control insects and fungal diseases respectively can be protectant or eradicant in action. The former types are applied before the pest or disease is present and the latter are applied when the attack commences. There are many pesticides with a broad spectrum of activity and some which are highly specific in their activity and available for a whole range of pests and diseases of vegetables.

The use of pesticides, however, does have other implications for crop quality. Some pesticides will remain in the product after harvest and these so-called residues can occasionally lead to quality problems. The level of residue may be unacceptable or the presence of the residue may lead to the development of undesirable taints or off-flavours in the processed produce, but these instances are rare in countries where good husbandary practices are employed.

When pesticides are used to protect crops during their growth or during storage after harvest, traces of the pesticide may remain on the produce and through to the finished product. Maximum Residue Limits (MRLs) have been

Figure 2.8 Spraying a pre-emergence herbicide on a crop of sown onions (courtesy Monsanto plc.)

set for individual pesticide active ingredients in specific products. The concept of setting an MRL is that if a pesticide is used as recommended, resulting residues in the crop will not exceed the MRL value. MRL values are set as high as is necessary to encompass recommended use but are not set unnecessarily high. It is important to realise that the MRL is a trading standard and not a safety standard.

Most uses of pesticides do not leave detectable residues in crops at harvest. Early season uses, that is, when pesticides are used in the early stages of crop development, for example, herbicides applied pre-planting and pre-emergence, tend to fall into this category. It is the later season and post-harvest uses, mainly of insecticides and pesticides, where the edible part of the crop is exposed to the agrochemical, which tend to leave the highest residue levels.

MRLs for pesticides in vegetables are now set by both international agencies and national governments. The Codex Alimentarius Commission, an international body set up jointly by the United Nations Food and Agriculture Organisation and the World Health Organisation, recommends MRLs for pesticides to member countries where their use gives rise to residues in commodities moving in international trade and where these residues are or may be a matter of public health concern. There are currently well over 80 pesticide active ingredients for which Codex MRLs have been set, and almost an equal number are under consideration. These MRLs are not legally binding but individual states generally endorse them. Individual countries such as the UK government may set statutory MRLs for use in that country. These may be accepted Codex international limits or unique if no international limit exists or if there are strong local reasons for setting a different level.

Taints or off-flavours caused by the use of a pesticide are a different problem. Here the pesticide is causing an undesirable flavour to develop in the product and occasionally has been most evident in processed vegetables.

Only a few pesticides have been shown to be capable of tainting food crops. Probably the best example is still hexachlorocyclohexane (HCH), which caused taints in crops such as carrots and potatoes, when these crops were planted in land treated up to ten years previously. Thiram and captan have also caused taints in some processed products. Malathion has been associated with 'catty' taints but only by stocks in which the container had been opened and left for a year, whereas malathion of recent manufacture caused no taints. The chemical which caused these taints was identified as mesityl oxide, which reacts with hydrogen sulphide to give the catty odour.

The absence of taints from the use of pesticides in processed foods in recent years is perhaps an indication of the rigorous testing that pesticides now undergo. In the UK most pesticides used on crops for processing are now tested for their potential to cause taints. Many processors that contract crops issue lists of pesticides that may be used and these are known not to cause taint problems.

2.5.4 *Harvesting*

Most vegetable crops for processing are harvested by machine, although for certain crops this may involve some manual input. The increasing sophistication of modern crop-harvesting machinery has been a major development in recent years. There is no doubt that the introduction of mechanical harvesters has transformed the economics of producing some vegetable crops as well as making the production of these crops on a large scale a practical proposition. For example, green peas are a popular canned and frozen product and without an efficient harvesting method it would not be possible to meet the demand for a high-quality and reasonably priced product. To hand harvest peas on a large acreage, shell the pod and process them would be a monumental task and therefore the processed pea industry is highly dependent on the mechanical harvester for the provision of raw material.

It is interesting to note the development of pea-harvesting techniques. At one time pods would have been hand harvested in the field but gradually more sophisticated harvesting machinery has been introduced. The first major advance was the introduction of the haulm cutter which cut the plant at the base and placed the harvested plants in a swath or windrow in the field. This could then be gathered up and taken to the next major introduction, the viner. This was at first a static machine often located at the factory which shelled the peas from the pod. The next development was the introduction of the mobile viner which could be taken to the field. These machines were able to pick up the

Figure 2.9 An FMC self-propelled pea harvester and viner (courtesy FMC Corporation (UK) Ltd).

cut plants directly from the swath and vine them on the move. This of course still required two operations and machines, swathing and vining. The modern generation of machines combines both operations in one pass (Figure 2.9). They have been developed for greater speed and capacity, to cause less damage to the peas, to give a cleaner sample and to give a greater recovery of peas.

As well as peas, French beans, broad beans, Brussels sprouts, tomatoes, carrots and potatoes can all be harvested by machine. Crops such as cauliflower and cabbage are still harvested by hand although harvesting aids are increasingly being used. Such aids comprise a tractor-drawn or self-propelled inspection area which is fed by arms of conveyors on to which workers place the cut heads as they work through the crop behind the machine. Some other harvesting operations involve manual sorting or inspection by a team on a platform which is fed by the mechanical harvester. Potatoes and tomatoes are cases in point (Figure 2.10).

The use of harvesting machines does have implications for quality. Crops must be grown to accommodate machine harvesting and different types of machine may have different requirements. Some cultivars are more suitable for mechanical harvesting and for French beans this is an important characteristic. The machine must also be set-up correctly, if not, problems can

Figure 2.10 Mechanical harvesting and hand sorting of potatoes in the field (courtesy Potato Marketing Board).

occur which result in damage to the produce. This may result in bruising, breakage, scuffing or cutting which are all important quality defects; it may also result in the admixture of stones and soil in the harvested produce.

2.5.5 *Storage*

In general, once the crop is harvested, quality begins to deteriorate. Either it is processed as soon as possible in some form or other to arrest deterioration or it is placed in conditions which will considerably slow the rate of deterioration. It is important, therefore, that if the harvested produce is to be held for any length of time before processing, there is no undue loss of quality.

Vegetables are commonly stored for periods ranging from a few weeks to many months or even years. Some vegetables will remain suitable for processing for long periods whereas others have a more limited post-harvest life. In general, those crops harvested immaturely during a rapid growth phase are the most perishable. Products such as green peas, French beans and cauliflowers have a very short shelf life and can only be kept for a few hours. These crops continue to respire at a rapid rate and, in addition for peas, the damage caused by mechanical harvesting increases enzyme activity which if not arrested quickly will lead to the development of off-flavours in the canned or frozen product. Cooling of the produce can slow the rate of deterioration but the period in which they can be held is still relatively short. Mushrooms are another highly perishable product which has to be processed within a short time of harvest.

Durable produce such as fully mature peas and beans can be kept for years providing they are stored with a relatively low moisture content. More perishable but nonetheless reasonably durable vegetables such as potatoes, onions and carrots can be kept from one season to another and Dutch white cabbage can be stored for many months, considerably extending the period of utilisation. The principle requirements for all these crops are the control of temperature and relative humidity. These factors are often critical and commodities have to be kept under different conditions depending on the intended use. Potatoes stored for processing, for example, normally have to be stored at a higher temperature than ware tubers such as those destined for the fresh market, to minimise the effect of 'low-temperature sweetening'. The normal recommended holding temperature for potatoes for processing is 8–10 °C. At lower storage temperatures reducing sugars accumulate particularly during prolonged storage. The colour of processed products such as crisps and French fries is mainly dependent on reducing sugar content of the potatoes used for their manufacture, and too high a reducing sugar content will result in an unacceptable dark brown colour (Figures 2.11 and 2.12).

Although sophisticated methods of storage are available and are becoming more so every year, less sophisticated options are sometimes as good. Carrots, for example, can be stored *in situ* in the field simply by being defoliated and covered by straw or polythene.

As the quality of the produce can only, at best, be maintained after harvest, what goes into store is of utmost importance. This means that only high-quality produce should be considered for storage. If poor-quality produce is stored deterioration may be even more rapid. All in all there seems little point in storing produce which is either poor in quality or unsuitable for storage. Of equal importance to the state of the produce going into store are the conditions of storage. Unless the correct conditions are maintained the quality of the produce will rapidly deteriorate.

If a crop is destined for storage then decisions about storage will have to be made at an early stage. Some cultivars, for example, store better than others. The crop should also be harvested at the most appropriate stage of maturity for storage. The produce should also be free from pests and diseases and undamaged in the harvesting operations. Stores should also be well maintained and clean, there should be no debris from previous crops which may harbour pests and diseases.

It may be necessary to prepare the crop for storage. Root crops may need cleaning and some vegetables may need treating with pesticides to protect against storage diseases. Other crops may be treated with growth regulators to control secondary growth in store. For example, potatoes and onions are treated with sprout suppressants either before or during storage. Once in store, it may be necessary to condition the produce. Potatoes, for example, may be

Figure 2.11 Loading a bulk potato store (courtesy Potato Marketing Board).

given a short period of a higher temperature to encourage the healing of damage to tubers caused by harvesting and other handling operations. Onions may also be 'ripened' with a period of higher temperature which improves the skin finish. Both procedures have to be done carefully as they can lead to additional disease problems. For example bacterial soft rots of potatoes and mould such as those caused by *Aspergillus* spp of onions.

2.6 Aspects of production of some major vegetable crops

2.6.1 *Peas*

Peas are a very popular product and most are consumed in a processed form. The majority of peas are canned or frozen although some are dehydrated

Figure 2.12 Loading a boxed potato store (courtesy Potato Marketing Board).

(specially for soups). In the UK more peas are processed than any other vegetable, with the possible exception of potatoes.

Peas for canning and freezing can be divided into a number of categories. They may be processed fully mature or immature. Fully mature peas are ripened naturally on the plant and harvested dry. Ideally the moisture content at harvest is no more than 18% and they may be stored for long periods and utilised at any time of the year. When peas are harvested at a higher moisture content they have to be artificially dried. Two types of pea are canned. The small blue/Alaska type is a small round-seeded pea. The second major type is the marrowfat, which is wrinkle-seeded and larger than the round-seeded Alaska. There are several cultivars of both types which are preferred for canning. The best quality Alaska type peas are produced in the Palouse area of Idaho and Washington in North America.

Immature peas may also be divided into two major groups. One group is round seeded. The seeds are usually small and pale seeded and at harvest have a relatively high level of starch. They are more suitable for canning than

Table 2.2 Typical size grades of immature green peas.

Size grade	Large seeded cultivars (mm)	Petit pois or small seeded cultivars (mm)
Large	> 10.2	> 9.3
Medium	10.2–8.75	9.3–8.75
Small	8.75–7.5	8.75–8.2
Very small	< 7.5	8.2–7.5
Extra small	—	< 7.5

freezing and are used widely in mainland Europe. The second group is the wrinkle-seeded type where seeds tend to have a higher sugar content and hence are naturally sweet. These peas tend to be larger seeded although some newer cultivars have small seed size. Cultivars of wrinkle-seeded peas may be deep green or pale green. The latter tend to be used for canning only whereas the deep green cultivars may be frozen or canned. Canned and frozen peas in the UK are mostly of the wrinkle-seeded types.

The size of pea is an important quality characteristic and smaller peas are increasingly in demand. Peas may be classified according to their size (Table 2.2) and plant breeders have been successful in introducing cultivars which yield well in the smaller size grades. Peas may be graded after harvest and the different sizes packed separately (as is practised in mainland Europe) or the peas may be grown specifically by careful choice of cultivar for a particular market.

Choice of cultivar is, therefore, very important for peas depending on whether they are for harvesting immature or mature and for freezing or canning. The other important aspect of production, which has already been discussed in detail, is the determination of the correct time of harvest.

Whole immature pea pods may also be processed. Although the market for these is much smaller, some are frozen. Two types are common; the 'Mange Tout' and 'Snap Pea'. Harvesting is the main problem and this is currently an area of research by processors.

2.6.2 *Carrots*

The carrot is an important root vegetable for processing and considerable amounts are canned and frozen. Unlike peas, it is also an important fresh vegetable. Different groups of cultivars are required for the different uses. Carrots can be broadly classified by their root shape (Figure 2.13). In general, cultivars of carrots fall into one of five categories, although a recent trend has been for plant breeders to introduce cultivars which are a cross between parents from different groups. These incorporate the best characters of both groups.

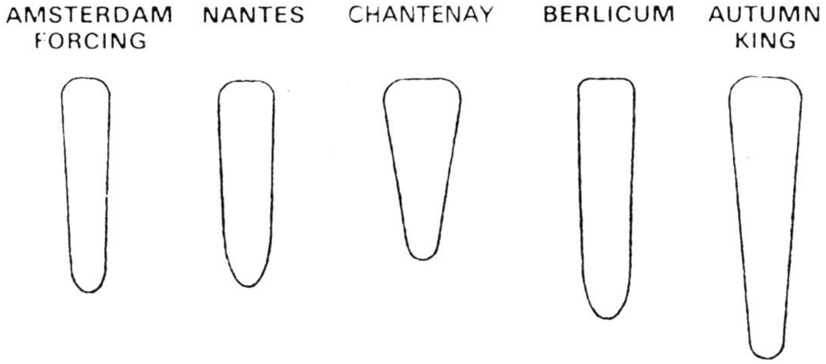

Figure 2.13 Typical root shape of carrots (courtesy the National Institute of Agricultural Botany).

In the UK the Chantenay type has traditionally been grown for canning whole while the more cylindrical Amsterdam Forcing types have been grown for freezing although they are increasingly being used for canning. For slicing and dicing the larger Autumn King and Berlicum types are grown for both freezing and canning. In general, small roots are required for processing whole and the larger ones are more suitable for slicing and dicing.

Carrots require careful husbandry during production in the field. The seed rate is very important to establish the correct plant population because on this the size of roots at harvest depends. Consequently seed bed preparation is important as this will influence establishment and ultimately plant population. For the same reason only seed of known quality and graded size should be sown. There are a number of important pests and diseases of carrots which influence processing quality and therefore must be controlled. Carrot root fly is a major pest and can be difficult to keep in check. Cavity spot is a major disease which is also difficult to control. It is the most serious defect in carrots for processing because even after steam peeling the lesions are still present. The cause of this problem was for a long time unknown and was most often classified as a disorder associated with calcium deficiency. Recently, however, the causal agent has been identified as a fungus and attributed to different species of *Pythium*. As a result, specific control measures have been introduced involving the use of fungicides active against *Pythium* spp.

2.6.3 *Beans*

Like peas, beans can be broadly classified by maturity into dry beans and immature seeds and pods. Unlike peas, the market for whole immature pods is as important as that of the separated seeds.

A number of types of dry bean are canned including the navy bean and dark red kidney bean and many other types are also processed but on a smaller scale. The processed white navy bean or pea bean is an important product in the UK where it is canned in tomato sauce. These beans are primarily grown in North America where conditions are most suitable for the cultivars currently available. Plant breeders have endeavoured to breed cultivars suited to European conditions but the quality cannot yet match those grown in the United States and Canada because of climatic conditions. The beans are grown to full maturity and at harvest the dry plants are cut and formed into swaths early in the morning while the dew is still on them to prevent shattering of the pods. The swath is then picked up by a combine harvester later in the day when the plants have dried out.

Green beans, which consist of the whole pod, are harvested immature and are grown for freezing and canning. Traditionally this type of bean is a climber but modern varieties produce a bush-like plant from which the pods are harvested mechanically. If it were not for this advance by the plant breeder it would be impractical to produce this crop for processing. Because all beans for processing are now harvested mechanically the cultivars used must have certain characteristics which render them suitable. These characteristics include uniform maturity of pods, an upright plant and one that is well anchored to the ground as the pods are combed from the plant by the machine.

Like peas, the immature seeds of beans may be used for processing. Both green beans (or flageolets) and broad beans are grown for canning and freezing. In the UK canned broad beans should have a pale green colour. This colour in the processed product can only be obtained by using white flowered cultivars. Beans from coloured flowered cultivars contain leucoanthocyanins which, after heat processing, convert to coloured polyphenols which give the bean a reddish-brown colour which is not acceptable to the UK market although it is acceptable in other European countries. Both white and coloured flowered varieties may be frozen but the colour of the bean remains pale green.

2.6.4 *Brassicas*

Brassicas grow particularly well in cooler, wet climates and are well suited to conditions in the UK and northern Europe. The most important brassicas for processing are cabbage, cauliflowers, broccoli or calabrese and Brussels sprouts.

Cabbages are widely grown and different types are available all year round. The fresh market is the most important but cabbages may also be processed. Shredded cabbage from headed types may be frozen or fermented to produce sauerkraut which may be canned or packed in glass. Special cultivars are grown for sauerkraut production and cabbages for processing are often planted at low populations in order to produce large heads which are more easy to handle. Some headed cabbages like Dutch whites may be stored for several months and, therefore, available for use over a long period. They have to be stored at low temperatures (1 °C) and treated with a pesticide as spoilage diseases can cause considerable losses in store.

Brussels sprouts are widely grown for freezing but may also be canned, although only rarely in the UK. For freezing, small, tight green sprouts are required whereas the fresh market requires a larger sprout. Most Brussels sprout crops for processing are partially mechanically harvested in one operation and those for the fresh market may be hand picked over several times. Because of the requirement for a single harvest crop for processing the plants must be grown in such a way that all the sprouts on the stems are encouraged to mature together. This is achieved by removing the growing point when the sprouts at the base of the stalk have reached 12 mm in diameter. Alternatively, modern cultivars have been developed with uniform maturity. Brussels sprout crops for processing are also established at relatively high plant populations to ensure that a high proportion of the sprouts are in the smaller size grades.

Cauliflower and broccoli (or calabrese) are increasingly popular frozen products. Cauliflowers are a specialist crop to grow, requiring the right soil type and good management to produce a high-quality crop for processing. The heads must be carefully selected in the field for quality and the right stage of maturity for harvest. Unlike Brussels sprouts, cauliflower crops may be harvested over a period of time even though modern cultivars are bred for uniform maturity. To a certain extent the grower can manipulate maturity and recently sophisticated methods for predicting maturity have been developed. These techniques together with the selection of cultivars of different maturity can be used to give a continuity of supply. Calabrese is a similar specialist crop. Interest in this crop for processing is more recent and husbandry techniques are less well developed. Breeding of cultivars specifically for this market is also in its infancy.

2.6.5 *Potatoes*

Potatoes are grown for many purposes, each market having its own requirements and needing a different quality of tuber. As a result several distinct types of production exist. The main markets for processing are canned potatoes, potato crisps, frozen potato chips or French fries and dehydrated potato products. For canned new potatoes the plants should give a high yield of small uniform tubers between 20 and 40 mm with cream or white flesh, firm waxy texture and dry matter under 20%. Canned diced potatoes require large 'blocky' tubers over 50 mm, white flesh, waxy texture and dry matter below 21%. Crisping (or chipping) potatoes should be uniform, with regular shaped, medium to large sized tubers over 40 mm with dry matter over 21% and reducing sugar content below 0.2% (preferably below 0.1%). French fry products require medium to large regular shaped tubers over 45 mm with

Figure 2.14 Sorting diseased and damaged potatoes after peeling (courtesy Frigoscandia).

cream or white flesh, dry matter over 20% and reducing sugar content below 0.25% (Figure 2.14). Dehydrated potatoes as flakes and granules require evenly shaped tubers over 25 mm with no greening, a cream or white flesh, dry matter over 20% and reducing sugar content under 0.5%. They are sometimes produced from off-cuts from other potato-processing operations.

Potatoes are influenced by production factors more than any other vegetable. Factors such as choice of cultivar, maturity at harvest, soil type and site, husbandry methods and storage conditions all have a profound influence on suitability for a particular use. Canned new whole potatoes, for example, are produced from relatively few cultivars. Higher than average seed rates with close planting densities are required. Multisprouting of the seed is also encouraged to induce more main stems which produce a larger number of small tubers. Haulm destruction is carried out to stop the dry matter from reaching too high a level. Potatoes for frozen French fry production, on the other hand, are planted at lower densities in order to produce a high proportion of large tubers. The crop is also grown to reach maximum maturity to increase dry matter and give a low reducing sugar content. Cultivars specific to French fry production only are grown. Storage of potatoes for processing is also important, although new potatoes for canning are generally utilised within a relatively short period following harvest. In general, potatoes for processing are stored at relatively high temperatures (8–10 °C). As observed earlier, if lower temperatures are used the reducing sugar content may become too high. Chemical sprout inhibitors also have to be used as potatoes will sprout naturally after about January or February. In general, the higher the storage temperature the shorter the dormant period, that is, before the development of sprouts. Sprouting is undesirable as it leads to changes in the quality of the tubers and is associated with an increase in reducing sugars.

Further reading

1. Anthistle, M.J., 'The composition of peas in relation to texture', *Scientific Bulletin* No. 4. Fruit Veg. Cann. Quick Freez. Res. Ass., Chipping Campden, Glos., England, 1961.
2. Arthey, V.D., *Quality of Horticultural Products*, Butterworths, London, 1975.
3. Arthey, V.D., 'Quality control of fruits and vegetables and their products', In *Quality Control in the Food Industry*, Vol. III (2nd edn), Academic Press, London, 1986.
4. Bassett, M.J., *Breeding Vegetable Crops*, AVI, Westport, CT, 1986.
5. Biddle, A.J. and Knott, C.M., *Pea Growing Handbook*, Processors and Growers Research Organisation, Peterborough, 1988.
6. Blund, B.F., *Crop Production: Cereals and Legumes*, Academic Press, London, 1971.
7. Bould, C. and Hewitt, E.J., *Diagnosis of Mineral Disorders in Plants*, Vol. 1: *Principles*, HMSO, London, 1983.
8. Burton, W.G., *The Potato* (3rd edn), Longman Scientific and Technical, London, 1989.
9. Blake, F., *The Handbook of Organic Husbandry*, The Crowood Press, Marlborough, 1987.
10. Dixon, G.R., *Vegetable Crop Diseases*, Macmillan, London, 1981.
11. Gane, A.J., King, J.M. and Gent, G.P., *Pea and Bean Growing Handbook*, Vol. II: *Beans*, Processors and Growers Research Organisation, Peterborough, 1975.
12. Halley, R.J. and Soffe, R.J., *The Agricultural Notebook* (18th edn), Butterworths, London, 1988.
13. Hebblethwaite, P.D., Heath, M.C. and Dawkins, T.C.K., *The Pea Crop*, Butterworths, London, 1985.
14. Holdsworth, S.D. *The Preservation of Fruit and Vegetable Food Products*, Macmillan, London, 1983.
15. Lyon, D.H., 'Taints arising from use of crop chemicals', in *Quality in Stored and Processed Fruit and Vegetables*, Academic Press, London, 1981.
16. Pattee, H.E., *Evaluation of Quality of Fruits and Vegetables*, AVI, Westport, CT, 1985.
17. Rastovoki, A. *et al.*, *Storage of Potatoes*, Pudoc Wageningen, 1987.
18. Scaife, A. And Turner, M., *Diagnosis of Mineral Disorders in Plants*, Vol. 3: *Vegetables*, HMSO, London, 1983.
19. Smith, I.M., *et al.*, *European Handbook of Plant Diseases*, Blackwell Scientific Publications, Oxford, 1988.
20. Walker, J.C., *Diseases of Vegetable Crops*, McGraw-Hill, New York, 1952.
21. Wills, R.B.H. *et al.*, *Postharvest. An Introduction to the Physiology and Handling of Fruit and Vegetables*, BSP Professional Books, Oxford, 1989.

3 Preparation procedures

PETER RUTLEDGE

3.1 Introduction

The preparation of vegetables for processing is most important if quality products are to be manufactured. The deterioration of vegetables starts at the time of harvest and this can only be minimised by correct handling and pre-processing techniques. These involve:

- Receipt at factory
- Conveying
- Cleaning
- Washing
- Inspecting
- Trimming
- Peeling
- Slicing or dicing
- Blanching

There are general methods of doing these operations but for some vegetables specialised machinery is required.

When designing and building a processing line for preparing vegetables, the major economic consideration is the cost of operation and the amount of time the equipment is being utilised. The manufacturing costs of the equipment in nearly all cases are secondary to the costs of operation. Operating costs on a per unit of vegetable processed basis take into account the type and use of energy, the labour employed, the amount and type of water used and the cost of effluent disposal.

The energy used is of two types. Electrical energy is used to run motors driving pumps and the mechanical moving parts of the machines and conveyors. The other significant energy source used in food preparation for processing is heat in the form of steam or hot water. Steam is produced in boilers fired by coal, gas, oil and in some cases, wood or waste material, depending on which is the most cost effective. The energy used for control systems is usually very minor.

Water usage is a major consideration in plant design and operation. The use of the same water for more than one operation is becoming more common as the cost of water purification increases. For example, it is not uncommon for cooling water from the blanching operation to be re-used as initial product washwater. Gallop (13) has advocated the closed-loop system of water usage where the same water is used many times. Local authorities charge for treating effluent received from factories according to the chemical and biological oxygen demand. These costs are increasing to the point where many factories are forced to treat their own effluent before discharge from the plant. These factors increase the demand for processes and equipment which reduce water usage and the biological oxygen demand (BOD) of effluent.

3.2 Receipt of vegetables into the factory

Vegetables are normally transported to the factory by truck or trailer. The load is weighed at the factory gate to determine both the payment to be made to the supplier and the amount of vegetables waiting to be processed. Samples of the vegetables are taken at this time to determine whether or not they are of the quality required by the factory. Factory standards for factors such as vegetable maturity, time and temperature of transport, extraneous vegetable matter, extraneous animal matter, amount of adhering soil, damage to the vegetable and presence of harmful material such as glass or metal must be complied with before receipt is granted.

Contracts between the supplier and the factory include the required factory standards. In some cases the factory will contract the growing of the crop, including the provision of seed, and determine all conditions from the time of planting to the harvesting and transport. In these cases the initial inspection will not be as thorough as when the grower contracts to supply cleaned, washed vegetable of processing maturity.

The time and temperature between harvest and blanching is most important for many vegetables to maintain quality. Bulk transport of mechanically harvested vegetables such as peas, beans and sweet corn present problems with self heating due to respiration and may require cooling before transport. The receipt of vegetables and the organisation of supplies in these cases must be efficient to prevent undue delays.

3.3 Conveying

The movement of product from one process to another involves conveying systems of different types depending on the vegetables and the processing line. It is important that conveying systems do not damage the product, for example, by having large drops from the conveyor into other machines. One criterion for selecting a conveyor design must be that it minimises the mechanical damage to the product.

3.3.1 *Flumes*

A flume is a trough containing running water. Vegetables are dropped into the water stream and transported by suspension until they are screened out at a dewatering station. Water is then pumped back to the start of the flume for recycling. The water acts as a cushion to prevent damage to the vegetables especially on corners. Riffle plates can be placed in the bottom of flumes to remove small stones and other dense objects. Flumes may also be used to cool vegetables after blanching and to transport the product to a freezer or can filler. The direction of flumes can be changed easily compared to most mechanical conveying systems.

The disadvantage of flumes is the leaching of solutes from the vegetables especially after blanching. Care must be taken to ensure that microbial build-up does not occur in the flume water which could contaminate product. Flume water is another source of high BOD effluent as it often contains relatively large amounts of soluble solids and finely divided insoluble solids.

3.3.2 *Pumped systems*

Vegetables are processed on a seasonal basis and different types of washing and preparation machinery can be situated at a distance from centralised operations for blanching and final processing. A pumped system is useful in those cases where product has to be transported relatively long distances within the factory. The vegetables are mixed with water on the inlet side of a specially vaned centrifugal pump. The mixture is pumped through pipework to a dewatering station where the vegetables are separated from the water. The water usually is returned to the pump by a gravity return line. The pumps are situated at a lower level than the vegetable outlet to facilitate the water return. The pumped systems require smooth bends for direction changes and they have the same problems as those associated with flumes in the treatment and disposal of water.

3.3.3 *Mechanical conveyors*

Mechanical conveyors can be categorised into belt or vibratory systems. The most common conveyor is for the inspection operation and consists of a rubber or plastic belt running on two drums, one of which is driven. The colour of inspection belts is important and should contrast with the vegetable to prevent eye strain. The tracking of the belt on the drums can be a mechanical maintenance problem. Smooth belts are used when transporting material in a horizontal plane but when the material requires elevation then cleated belts or bucket elevators are used. Other systems used for larger vegetables include sets of driven rollers. Roller systems turn the vegetables over as they are transported and are often used for product inspection. All the belt systems have hygiene problems in that vegetable soil lodges on the drive drums and is

Figure 3.1 A vibratory conveyor (courtesy Frigoscandia).

then squashed onto the underside of the belt contaminating all the supporting slides. Scrapers or water sprays are often used to remove product adhering to the return belt.

The vibratory conveyor (Figure 3.1) operates by an out-of-balance weight driven from an electric motor or from an electromagnetic vibrator. Plant material may be vibrated on a table which is slightly higher at the inlet end. The material is moved by gravity aided by vibration. Alternatively a method can be used which can move the material up slight inclines. Positive vibrations in the forward direction throw the vegetables forwards and upwards and while the vegetables are in flight the table retracts to the initial position. The effect of this is to move the product in the forward stroke but have the product slip on the table in the backward stroke. Vibratory conveyors are much easier to clean than belts and do not have the same hygiene problems. Vegetables that tend to mat, such as green beans, are shaken apart on vibratory conveyors. Such conveyors, however, are not used for product inspection because it is difficult to focus on the moving product.

3.4 Dry cleaning

Vegetables that are harvested from parts of the plant growing above the ground often have large amounts of extraneous vegetable material which must be removed before processing. A common method of achieving this is by air cleaning which involves passing a stream of high-velocity air through the product so that the lighter unwanted material is entrained in the air stream

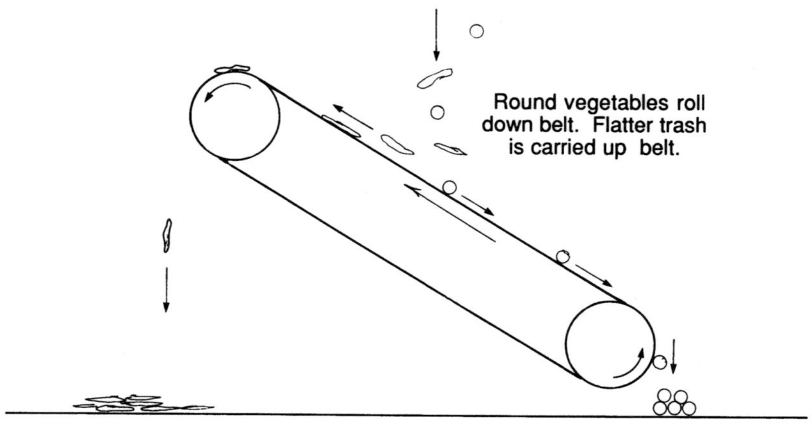

Figure 3.2 Inclined belt cleaner for round vegetables.

and removed. This method will be of use only where the vegetables are of greater density than the extraneous material. Mobile pea and bean harvesters are fitted with this type of cleaner, so most of the unwanted material remains in the field thus reducing the waste disposal problems at the factory.

Vegetables harvested from below the ground may be given an initial cleaning treatment to remove excessive dirt. Such vegetables are often passed through a series of rotating brushes which remove adhering soil. The soil is then returned to the field and does not create a problem in a factory's effluent system.

Some cleaning techniques rely on the shape of the vegetables to effect separation of the trash. For example round vegetables may be rolled down inclined ascending belts. The vegetables roll to the bottom and are removed and the extraneous material which will not roll is carried on the belt and discharged over the top as shown in Figure 3.2.

Dried pulses, such as navy beans and peas, sometimes have as contaminants volcanic porous stones of similar density and size to pulses. This material is difficult to remove but can be separated by vibrating the beans across a slightly inclined plate known as a gravity separator. The beans follow one contour and the stones another and are collected at separate outlets. Colour sorting machines are also used for removing extraneous material from dried pulses (see Section 3.6).

3.5 Washing

There are three general classes of vegetables: those grown under the soil such as root crops; leafy vegetables like spinach; and vegetable fruits and flowers such as peas, beans and cauliflower. The methods of washing differ, depending on the type of vegetable.

Rootcrops are often covered with soil which has to be removed. They are by nature much denser than water and will sink when placed in a tank of water. Therefore, if they are to be cleaned using an immersion system a carrier belt is required to move them through the tank. However, it is easier to use a rotary brush washer followed by rinsing in a rod washer (see Figure 3.3) to remove the soil from these vegetables. This operation does not have to be perfect as it is followed by skin removal in a steam or caustic peeling system. In many cases, if the soil is superficial, wet cleaning might not be required. Sometimes vegetables are conveyed into the factory in flumes and this action can be considered adequate washing when followed directly by a spray rinse and peeling.

Leafy vegetables create a washing problem as they layer when bulk stored. The washing operation must wash all surfaces including those within the layers. Leaves are washed by immersion in agitated water systems. A typical system is explained by Wright *et al.* (38) where the leaves are washed in water with high-pressure water jets and held under the surface with a paddle system. Other methods use compressed air jets in the water as the agitation medium to separate the leaves and wash all surfaces.

Green peas are cleaned on a vibratory grader to remove pods and peas that are undersize. The peas are then dry cleaned and washed in a rod washer (Figure 3.4) to remove 'viner juice'. Stones and heavy objects are removed in a float tank and in some factories froth flotation cleaners are used to remove seed heads, skins and other debris. Froth flotation cleaners were used by Veldhuis and Neubert (33) to remove nightshade berries from peas. This method relies on the ability of a surface active agent to cause air bubbles to

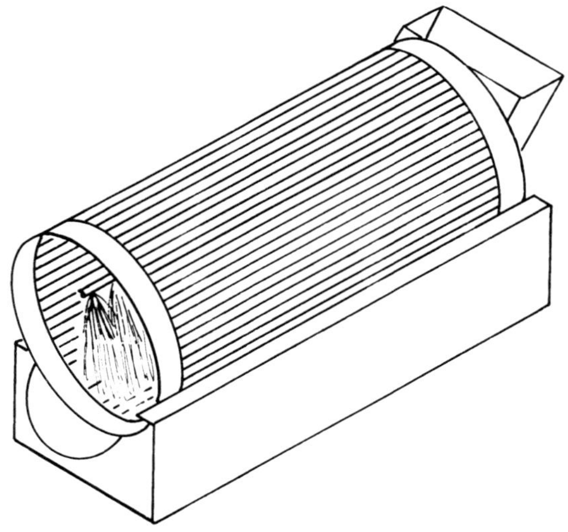

Figure 3.3 Rod washer with central spray bar.

Figure 3.4 Washing peas (courtesy Frigoscandia).

become attached to extraneous material which then can be floated away from the peas.

The methods used to wash tomatoes are different from most other systems. Tomatoes are relatively fragile so, on arrival at the factory, they are tipped into a tank of water in which they are suspended. The denser material such as dirt sinks, and leaves float free. The tomatoes are then elevated from the tank usually on a roller conveyor under water sprays.

3.6 Inspection

Manual inspection and picking of vegetables on inspection belts or roller sets is the traditional way of removing unwanted material from the production line (Figure 3.5). Where this system is used properly, it is the most labour-intensive operation in the factory. In factories operating in highly competitive markets, more economic higher speed vegetable processing lines are required. As this

Figure 3.5 Manual inspection of green beans (courtesy Frigoscandia).

evolves the inspection line becomes the rate limiting step. In some instances the inspection line becomes ineffectual being overloaded and understaffed; in others, staffing is increased to cope with the load and in some, methods other than manual labour are used.

There are several methods currently available for inspecting vegetables using the products of computer technology. These systems use detectors which reflect visible light, or infrared radiation, others use X-rays or television images. An optical sorting system for potato and carrot dice was reported by Somers *et al.* (31) which can detect defects of less than 2 mm^2 in area. Defective dice are removed from the stream using air jets. The machine inspects an average of 6000 kg of dice per hour and requires three operators where 40 inspectors were previously employed.

The use of a linear array camera by researchers at Essex University to detect incorrectly peeled potatoes was explained by Cowlin (8). The system is based on a television camera and monitor with the image being analysed by a computer system to identify defects. The system uses a presentation table with rollers to turn the product which runs at three times the speed of an equivalent manual inspection table.

Automatic Defect Removal (ADR) machines are so effective for the inspection of French fried potato strips that factories using them have virtually abandoned manual inspection of potatoes except to remove rotten vegetables from the line. Wilmont (37) reported that such machines can run at 12 000 kg of strips per hour. The strips are aligned into channels and scanned by a camera.

Defective strips can be removed either by air jets or in some machines the defect is cut from the strip with a retractable knife set in a drum above the channels. The defects are then separated on a size grader after the ADR unit.

The colour sorting of red and green tomatoes has been automated for many years with the scanner activating a flap at the end of the inspection belt to separate the two grades of vegetables. In the field the persistence of chlorophyll fluorescence can be used to separate ripe from green tomatoes. Navy beans have been optically sorted to remove small stones and stray seed material for many years.

The use of electronic systems for inspection of vegetables in general has made the inspection system faster, more accurate and eventually more economical. The advance of this technology is so rapid that by the year 2000 only a small fraction of the present level of manual inspection will still be carried out.

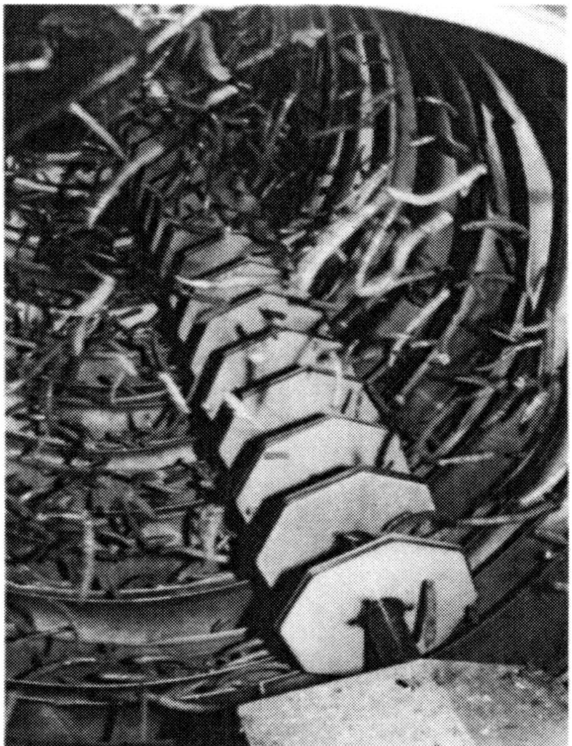

Figure 3.6 Declustering green beans (courtesy Frigoscandia).

3.7 Trimming

Some vegetables require trimming before they are processed. These trimming operations are similar to those carried out when the same vegetables are prepared in the home. For example, fibrous stalk is removed from asparagus, ends are snipped from green beans, stalks of cauliflower and broccoli florets are removed, and crowns are removed from carrot roots. Whereas a knife is used in the family kitchen for these operations, in the factory, specific machines are used to carry out most of these operations.

Asparagus is stacked with the crown end against a guide and cut to length using a fine-toothed circular saw or very sharp circular cutter. The fibre in the stem sometimes produces a ragged cut if the blades are not sharp. Care must be taken not to damage the relatively fragile crown.

Green beans pass through two systems to prepare them for processing. The first is a cluster buster which breaks the stems of beans that are joined in bunches on the plant and have not been separated during harvesting. The cluster buster contains sets of rotating cutting wheels to break the stems (Figure 3.6). The other specialised machine is a snipper to remove the ends of the beans. The beans are fed into a perforated rotating drum and the pods stand on end with the ends protruding through the perforations; cutters riding on the outside of the drum remove the protruding ends (Figure 3.7).

3.8 Peeling

The methods used for peeling vegetables such as carrots and potatoes are classified as mechanical, chemical and thermal.

Figure 3.7 Four green bean snippers, two with declustering wheels (courtesy Frigoscandia).

3.8.1 *Mechanical methods*

Machines that remove skin by rubbing action are commonly called abrasive peelers. These machines have many conformations but the most common type is the one using sets of abrasive rollers as in Figure 3.8.

The amount of skin removed is determined by the residence time in the peeler which can be adjusted by the speed of the auger in the centre. The rollers spin on their axes as well as revolving around the auger. This allows water sprays from the centre of the auger to clean those abrasive rollers not in contact with the vegetables. As this method abrades a plane surface only, crevices such as potato eyes must be removed by hand or the vegetable abraded until all the imperfections are worn away, which results in large peeling losses. Generally the yield of peeled product from abrasion peelers is lower than from other mechanical methods.

Peelers with knife blades mounted in the outside walls of a revolving drum in which the product is tumbled have a similar effect as an abrasion peeler. The difference between the two is the smoothness of the cut when sharp knives are used compared with the comparative roughness of the abrasion. These peelers, like the abrasion peeler, cut on a plane surface.

The material removed from mechanical peelers may be wet or dry depending on the vegetable and the type of machine used. Peeled skins are easily collected from the water stream in a wet system by straining, settling or centrifuging or by mechanical means if a dry system is in operation. The

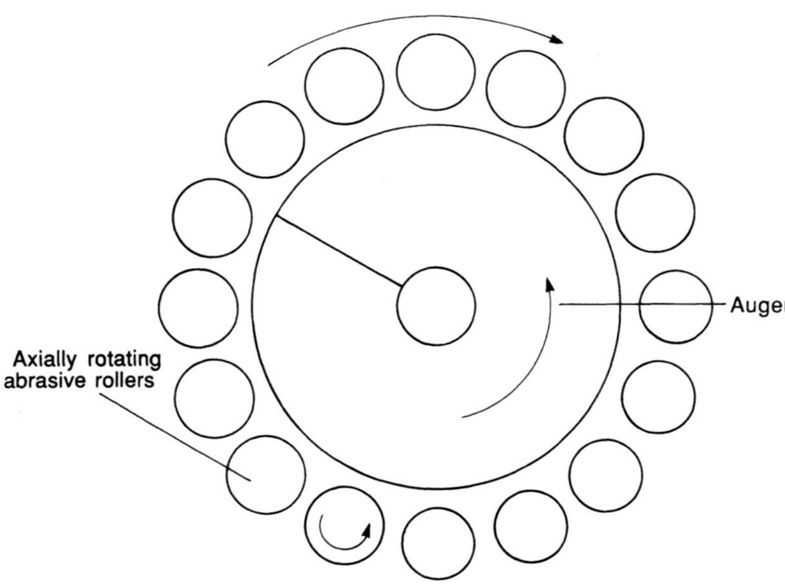

Figure 3.8 Abrasion peeler.

disposal of this material as a solid reduces the BOD of the factory's liquid effluent. This waste can be used as animal feed, fertiliser or for other by-products.

Some vegetable products require specialised peelers. For example, modern onion peelers clamp the bulbs in a machine so that the top and base can be removed by rotating cutters. A slit is made in the skin after topping and tailing and the onions are rotated past a set of high-pressure water spays which remove the skin (2). Sweet corn is dehusked by cutting the cob at each end and unrolling the cob from the leaves. The silks also are removed during this operation to leave a clean cob.

3.8.2 *Lye peeling*

The immersion of certain vegetables, such as roots, in hot sodium hydroxide solutions will effect peeling by chemical erosion of the skin and underlying tissue. The lye peeler comprises a bath of sodium hydroxide solution, a heat exchanger to heat the caustic solution and a means of transporting the vegetable through the bath. A paddle wheel system, as in Figure 3.9, ensures that the vegetables are totally immersed in the solution.

Peeling depends on the temperature of the solution, the concentation of the solution and the time of immersion. The addition of a wetting agent was recommended by Schultz and Smith (27) to increase the initial penetration of the caustic into the vegetable skin. The typical conditions for lye peeling some common vegetables are given in Table 3.1.

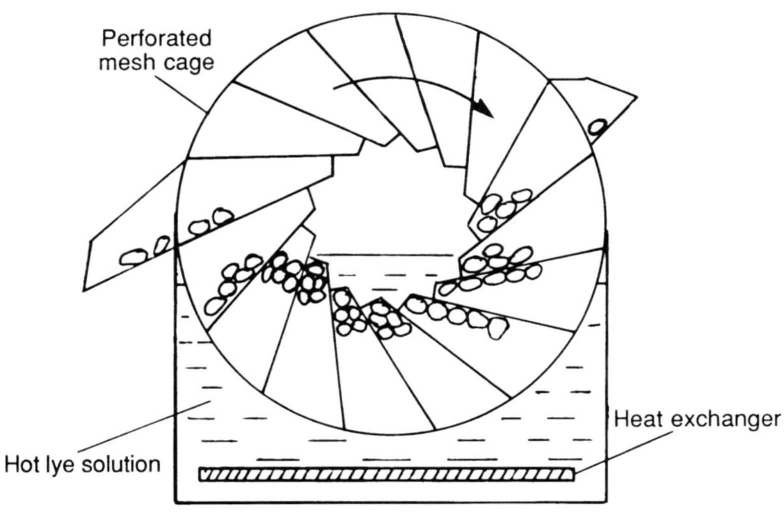

Figure 3.9 Lye peeler with perforated mesh cage.

Table 3.1 Lye peeling conditions for some vegetables.

Vegetable	Concentration (%)	Temperature (°C)	Time (min)
Carrots	5	95	1–3
Potatoes	8–18	60	2–7
Sweet potatoes	12	95	3–5
Onions	20	80	1–2
Tomatoes	16	90	0.5

Schultz and Smith (27)

Modifications to the standard lye treatment were recommended by Weaver *et al.* (34) for reducing the amount of lye used in peeling potatoes. A double-dip system was suggested where the potatoes were dipped in 4–12% lye solution for about 30 s at 93°C. They were then held for 5–15 min to allow the lye to penetrate the tissue of the skin. The potatoes were then dipped again in similar lye solution for 30–75 s and held for up to 3 min before removing the skin. The theory behind this method was to penetrate the skin with the first lye treatment and hold time and then penetrate the underlying tissue with the second lye treatment and short hold time.

Another system for peeling potatoes was outlined by Graham *et al.* (15) and Willard (36) where the vegetables were dipped in 17% lye solution at 75 °C for 5 min and then exposed to infrared radiation for 2–5 min before skin removal. The additional heat treatment helped to speed up the chemical reaction of the lye on the skin and tissue to effect the peeling operation. The result was a dry skin that could be brushed off before washing thus reducing the normal problems of disposal of wet, lye-saturated skins.

The disposal of lye treated skins is a major problem with this method of peeling. The use of potassium hydroxide instead of sodium hydroxide allows the dry or wet residue to be used as a fertiliser. There is a potential danger to plant operators when working with hot lye solutions and waterproof protective garments and eye protection should be worn when working with lye peelers.

3.8.3 *Hot water*

Hot water is used to prepare beetroot and tomatoes for peeling. Beetroot is cooked in near boiling water for about 30 min (depending on size) at which time the skin may be removed by gentle rubbing in a rubber-tipped scrubber. Tomatoes are scalded in boiling water for about 30 s which causes the skin to become loose and baggy looking. The tomatoes are then passed over a machine with many fine rollers which pinch the skin and drag it from the tomato. This is an alternative to lye peeling for tomatoes.

3.8.4 *Steam peeling*

When vegetables are held in high-pressure steam for a short time the skin and underlying tissue are superheated. When the pressure is suddenly reduced the tissue moisture boils instantly blowing the skin from the vegetable. This principle is used in steam peeling of root vegetables. The older type of peeler held the vegetables at about 6 atmospheres steam pressure for up to 2 min before releasing the pressure. Modern peelers use around 17 atmospheres steam pressure for 30 s. Steam peelers take many configurations with some being a vane type while others are sealed screw systems. High-pressure peelers tend to be batch types because of sealing difficulties with the higher pressures (Figure 3.10). Continuous steam peelers often suffer from wear on the mechanical seals due to soil particles carried with the vegetables. Steam peeling is the most efficient peeling system for potatoes and carrots. Some vegetables such as small potatoes and carrots for canning do not require blanching after steam peeling.

Figure 3.10 A batch type steam peeler in a potato processing plant (courtesy Frigoscandia).

3.8.5 *Skin removal*

With both lye and steam peelers, vegetables exit the machine with the skins loosened but not completely removed. Skin removal is most effective when the vegetables are tumbled in a high-pressure jet rotary washer. This removes the skin by the action of vegetables rubbing together and the water jets remove material from the crevices in the surfaces. The water also cools the vegetables and washes away the lye from the surface. The major disadvantage of this system is that it produces a significant effluent problem.

To combat the effluent from the cleaning operation two alternative machines have been constructed. One uses brushes to remove the peel as a solid as shown in Figure 3.11 and has been described in detail by Sijbring (30). The other uses rubber fingers mounted on rollers and has been described by Graham *et al.* (15). This type of machine removes most of the peel and is followed by a high-pressure water rotary washer to remove the remainder and to rinse the vegetables.

The peeling efficiency of the different methods of removing skin from vegetables is often compared to determine the best method. This is usually expressed as the percentage yield of peeled vegetables. The comparison is only valid if the vegetables are of the same size grade. Ratcliffe (22) argued that a ton of small potatoes had in excess of three times the area of skin as a ton of large potatoes and that the removal of the same depth of skin produced 67 and 91% yields of peeled potatoes respectively. The comparison of peeling methods from different sources does not, therefore, always give a real indication of

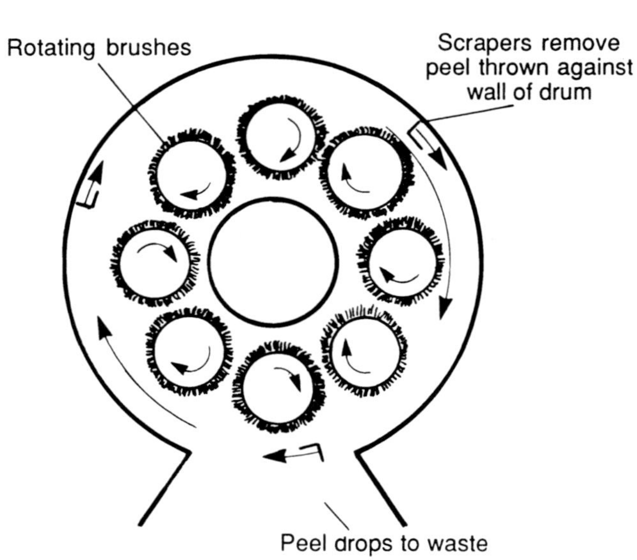

Figure 3.11 Dry peel removal using brushes (Sijbring (30)).

efficiency. However, on medium-sized potatoes the peeling loss for abrasion peeling is 20–25% and for steam and lye peeling 10–16%.

3.9 Slicing and dicing

Many vegetable products are not packed in the same form as they are harvested. At some stage in the processing, size reduction takes place to make the vegetable more acceptable to the consumer. There are many types of machines available to carry out size reduction although most can be categorised as slicer–dicers. The food industry relies on a few general shapes which are either slices straight, slices wrinkle cut, strips straight and wrinkle cut (Figure 3.12), or dice.

The machines for producing these shapes generally have three sets of knives of different type to perform each operation. The initial operation is to slice the vegetable forming a slab. This can be carried out with rotating straight blades or by pushing the vegetable past a paring blade. The slab is then cut into longitudinal strips to form an object similar to a French fry. A gang of rotating circular knives are often used for this cut, although in some machines the blade that cuts the slab contains both vertical and horizontal cutters to perform these tasks. The longitudinal slices are then cut into cubes if a dice is required. The blades to do this are mounted in a cage with the relative speeds of the cage and the slice determining the size of the dice. A typical system is shown in Figure 3.13. A swinging blade can also be used for this operation.

Water knives are a relatively new innovation where the vegetable is cut by a

Figure 3.12 A chipping machine for producing French fry strips (courtesy Frigoscandia).

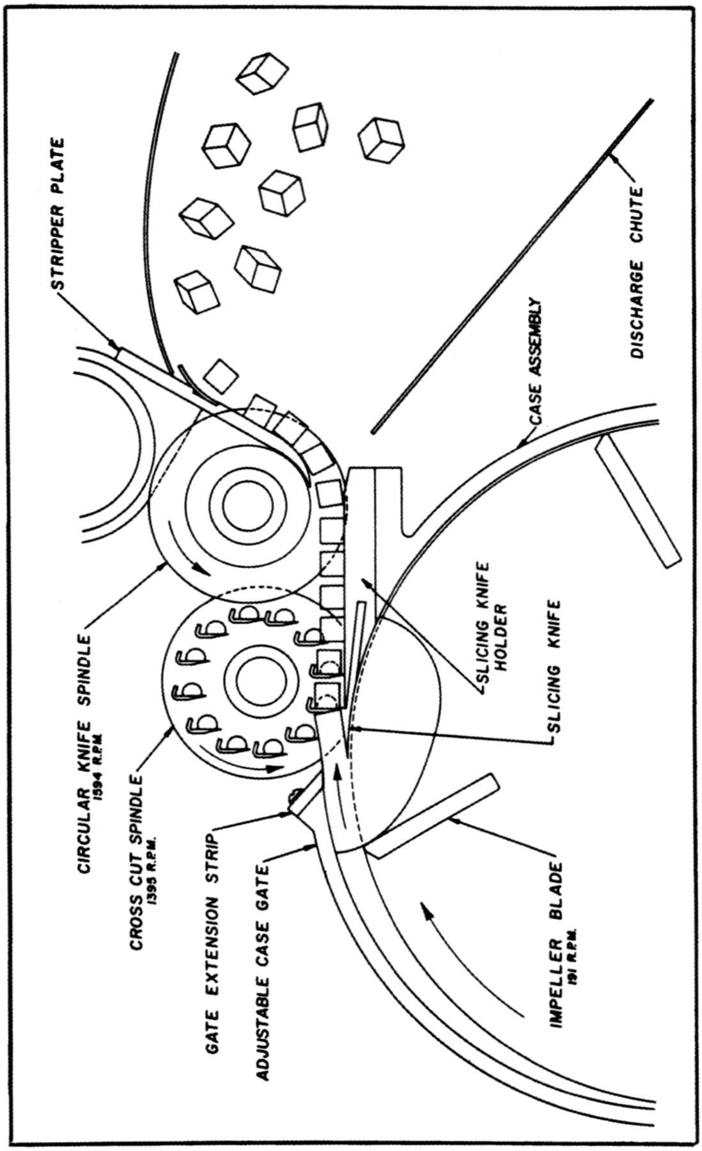

Figure 3.13 Method for dicing vegetables (courtesy Urschel Laboratories Inc).

Figure 3.14 A grading table for the removal of small pieces of potato from French fry strips (courtesy Frigoscandia).

fine jet of high-pressure water. The pressures can be up to 300 000 kPa producing a hair-thin stream of water through a sapphire orifice with a velocity of three times the speed of sound. The cut from these knives is extremely accurate and sharp. The water knife is non-directional as is a blade, and it has many applications where computer-controlled positioning is used (1).

Another water knife machine for cutting vegetables such as potatoes is explained by Wilmont (37). Whole potatoes are pumped in a water stream at high velocity through a pipe. Blades in the form of a grid are mounted on the outlet end of the pipe to cut the potato into French fries.

After the vegetable has been cut into slices or dice, there is always some trim material which needs to be removed. These are the pieces from around the edges of the vegetable which have not taken the squared shape. Such material is called slivers or nubbins and is removed on vibratory grading tables (Figure 3.14). In the case of vegetables such as potatoes a layer of released starch granules from the cut surfaces needs to be washed off if this material is not to degrade the blanch water. This material can usually be collected in a settling tank and in the case of potatoes has some commercial value as potato starch.

3.10 Blanching

Vegetable products are living parts of plants which continue to respire after harvest. When vegetables are harvested, deteriorative changes, often initiated by the plant's enzymes, start to break down the plant tissue. The time between harvesting and inactivation of these enzymes can be critical to the quality of the final product. This time delay is more important with actively growing leafy vegetables such as spinach, and crops such as green peas and beans, which may be damaged during harvesting, than with carrots and potatoes, which are derived from the plants' storage system. The juices from damaged tissue are an ideal substrate for microbiological growth. Microorganisms can cause heating and flavour changes if allowed to multiply to high numbers.

To prevent enzymic and microbial deterioration, the plant material is given a heat treatment which inactivates the enzymes and kills the plant tissue. This process is called blanching.

Blanching has other advantages depending upon the method of further processing the vegetables. Vegetables for canning are blanched mainly to shrink the vegetable to obtain the correct fill weight in the can and to remove the gases from the intercellular spaces which would otherwise cause oxidation of the product, corrosion of the can and low can vacuum. If there is a long delay between can filling and thermal processing then the inactivation of the enzymes will also be important.

Inactivation of enzymes is also important for preserving vegetables by freezing and drying. It prevents discoloration, softening and off-flavour development during subsequent storage. Blanching also has an additional cleaning effect and reduces the microbiological load of vegetative cells on the vegetable. Care must be taken with blanching systems to ensure there is no build-up of thermophilic bacteria as the temperature conditions in blanchers are suitable for their growth. Gillespy (14) surveyed flat-sour bacteria contamination in blanchers and found that although the blancher water temperatures prevented their growth other surfaces in the blanchers presented ideal conditions.

The enzymes that are the most heat resistant and can be used as indicators for adequate blanching of vegetables are peroxidase and catalase. Williams *et al.* (35) investigated blanching indicator enzymes for frozen vegetable production. They reported that the inactivation of catalase required only 50–70% of the heat treatment required to inactivate peroxidase. Also the deterioration of unblanched frozen vegetable was not due to peroxidase but mainly to lipoxygenase. They concluded that the heat treatment could be reduced to the point where lipoxygenase was inactivated particularly for green beans.

The commercial blanching process heats the vegetable as quickly as possible to the required temperature and then rapidly cools it to ambient. Rapid heating and cooling is used to minimise the softening of the tissue due to thermal damage. However, for some vegetables a long low-temperature blanching treatment is used to give the required texture.

The machinery for blanching vegetables has been modified in recent years to reduce the cost of energy used in processing and the amount of effluent which has to be treated. Steam and near boiling water are generally used as the heating media and hence blanchers are categorised into those utilising water and steam.

3.10.1 *Water blanching*

This is the traditional form of blanching which involves the product being held in hot (85–100 °C) water until the enzymes are inactivated, and then water cooled (Figure 3.15). The most common form of machinery for water blanching is the reel blancher which, in its simplest form, is a screw conveyor in a trough filled with hot water. The vegetables are carried through the water on the screw conveyor, which controls the retention time, and then water cooled.

The energy efficiency of the steam injection system used in many of the early reel blanchers was only 30–35% as a large proportion of the steam injected was lost to the atmosphere. Rumsey *et al.* (26) replaced the steam injection system with an enclosed heat exchanger and the efficiency of the blancher increased to 67%.

Reel blanchers are also used to cook product as in the case of potato slices for frozen French fries where two blanchers are often used in series. Using this system the texture required in the chips and an adequate production rate are maintained. These blanchers operate with a water temperature of about 85 °C. Colour of the fried chips is largely dependent on the amount of reducing sugar near the surface of the tissue. Blanching leaches reducing sugar to produce even coloured chips and in instances where the fried chips are too light reducing sugar may be added to the blanch water of the second blancher or the chips run through a solution containing reducing sugar after blanching.

A pipe blancher is sometimes used for final rehydration and blanching of

Figure 3.15 Hot-water blancher and cooler (courtesy Frigoscandia).

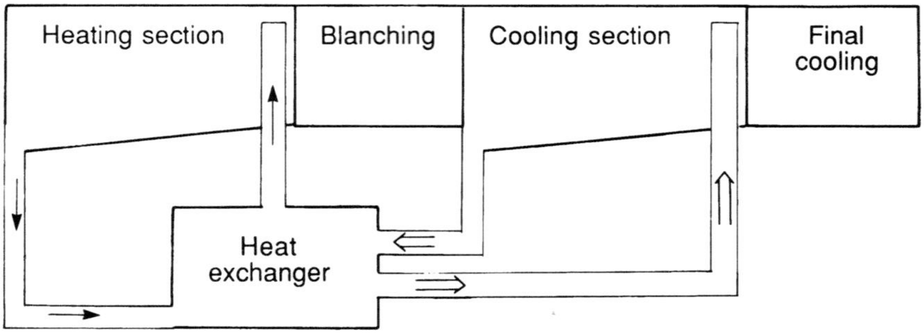

Figure 3.16 Layout of blancher using heat regeneration.

navy beans and dried peas. The product, suspended in water, is pumped through heated and cooled pipes. Rehydration using this process can take up to 20 min so the length of pipe required is quite extensive. In most cases steam is injected into the pipe at several points in the heating section. The advantage of this continuous system is that temperatures above 100 °C can be achieved to speed up the rehydration process.

The potential for regeneration of heat in a water blancher was reported by Byrne (5) in Denmark. The water in the cooling section was used to heat the water for a pre-heat section and vice versa as shown in Figure 3.16. Both water streams were filtered to remove fragmented solids before passing through the heat exchanger. The blanching section was heated with steam and recirculated condensate. The soluble solids leached from the vegetables build up in the recirculated water streams forming an isotonic solution. When the concentration of soluble solids in the water approaches that of the vegetable then less leaching takes place reducing the overall effluent and increasing the nutrient content of the blanched vegetable.

3.10.2 *Steam blanching*

The main advantage of steam blanching is to reduce leaching of solutes from the vegetable. This improves retention of soluble nutrients and reduces the effluent from the blanching operation. Vegetables that are normally steam blanched include cut sweet corn and broccoli. These products have exposed cut surfaces that are easily leached by water blanching.

Steam blanchers that consist of a mesh belt travelling through a steam chest or tunnel are generally being replaced by more efficiently sealed steam blanchers. The loss of steam at the product inlet and exit for the tunnel blancher is considerable and the design of the steam chest or tunnel often makes them difficult to clean. A blancher that used 50% less energy than the conventional blancher was described by Ray (24). This blancher is similar to the conventional belt blancher except that the product entry and discharge is

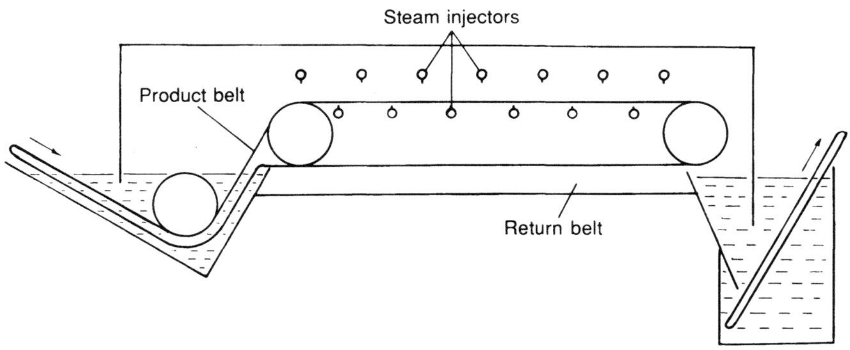

Figure 3.17 Water-sealed steam blancher.

through a water seal as shown in Figure 3.17. The removable cover is also water sealed and is lifted on a gantry to allow the blancher to be easily cleaned.

A concept in blanching called the individual quick blanch (IQB) was proposed by Lazar *et al.* (19). In this system pieces of vegetable are spread in a thin layer on a rapidly moving mesh belt passing through a steam chest. Because there is only a single layer of vegetable the heating is unobstructed and therefore rapid compared to other systems where a bed of vegetable pieces several layers thick is heated. The pieces are only steam heated long enough to inactivate the enzymes half way to the centre of the piece. The pieces are then deposited on to a slower moving belt where the thicker bed of vegetable is allowed to thermally equilibrate before cooling as in Figure 3.18. Refinement of this idea has led to the development of two different commercial blanchers.

One of these designs was developed by Cumming *et al.* (9) of Agriculture Canada and has vane type inlet and outlet ports to the steam chamber. The

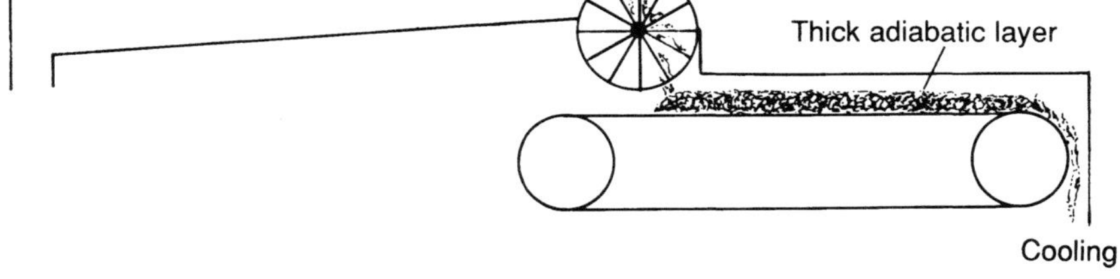

Figure 3.18 IQB blancher.

product, such as garden peas, is heated in a thin layer on a belt with saturated steam which is directed up throught the belt to impart a fluidising action on the peas. The peas are heated for 35 s before passing through the exit port and falling into a bucketed belt where they are held for a further 55 s before cooling. The 35 s blanch in steam is not sufficient to totally inactivate the enzymes as the heat does not penetrate to the centre of the peas. In the hold section the temperature in the peas equilibrates to give an adequate blanch. The claim by Dunlop (11) is that this blancher needs 90% less steam and produces 90% less effluent than a water blancher.

A vibratory product transmission system was used in another IQB blancher described by Brown *et al.* (4). This blancher transmitted the product down a spiral vibratory conveyor in the steam section and then held the vegetable in an insulated adiabatic holder until the blanch was completed. The vegetable then passed to another spiral vibratory conveyor where it was air cooled. A novel feature of this blancher was the return of the condensate from the steam section into the air stream of the cooling section as an atomised spray. This cooled the product and resulted in virtually no effluent being produced from the blanching operation. A full-size production machine was described (3) which would blanch 10 t of vegetable per hour.

A high-temperature short-time blancher was described by Robe (25). This machine was similar to a high-pressure steam peeler. The vegetables were loaded into the vessel and steam treated at pressure for about 90 s as the vessel rotated. They were partially cooled in the vessel then discharged for final cooling. The enclosed system used steam efficiently with a ratio of 8:1 for product to steam.

3.10.3 *Vacuum steam blanching*

Vacuum steam blanching is usually carried out in an end-over-end conical jacketed cooker. The vegetables are loaded into the cooker and a slow rotation started as the pressure is reduced to about 10 kPa. The vacuum is broken with saturated steam and the vessel brought to atmospheric pressure which is held for about one minute. The vegetables are then vacuum cooled by again reducing the pressure in the vessel. The liquid in the vessel flash boils under the reduced pressure cooling the vegetables.

The advantage of vacuum steam blanching is the rapid heating of the vegetable by condensation of the steam and hence liberation of the latent heat of condensation in the tissue. Because of the high-temperature short-time blanch, this produces firmer textured blanched vegetables, resulting in superior frozen products. The system is energy and effluent efficient. This type of blanching is also used commercially in the mushroom industry where the shrinkage of the blanched mushroom is not as great as with water blanching. The system is, however, only a batch operation and a good vacuum pump is required to achieve the low pressure.

3.10.4 *In-can blanching*

Mitchell (21) devised a system for water blanching of peas in the can. The method used was to overfill the can with washed peas and then fill the can with near boiling water. A jet of near boiling water was used to swirl the peas around the can to give good heat transfer. After about 30 s the peas had shrunk to fill the can and sufficient gas had been removed to give an adequate vacuum for canning. This process has not been used commercially.

A blanching-exhausting process was investigated by Casimir and Lewis (7) for vacuum-packed vegetables. The raw vegetable was filled into the can, a small amount of brine added (50 ml for a 400 g can) and the can lid clinched. The can in the upright position was passed on a conveyor over a ribbon burner. The brine boiled displacing the air in the can with steam and blanching the vegetable. The can was then seamed and thermally processed. The pressure in the can after this procedure was 20–25 kPa giving a good vacuum-packed product. This method of blanching is easily adapted to conventional processing lines and will run at normal vegetable canning speeds. The advantages of vacuum packing by this method are no blanching effluent and a considerable reduction in the leaching of nutrients from the vegetable to the reduced volume of canning brine.

3.10.5 *Microwave blanching*

Many workers have investigated the use of microwaves to blanch vegetables for processing. Decareau (10) lists blanching as one of the applications of microwave power in the food industry with corn, potatoes and fruit as examples, although there were no commercial applications at that time. The results of reported work differ widely, mainly because the microwave effect is not as easy to quantify as methods using a tank of boiling water or a steam cabinet.

3.10.6 *Hot-gas blanching*

Ralls *et al.* (23) investigated hot-gas blanching both at the pilot plant and commercial plant levels. The vegetables were heated with a mixture of steam and flue gases from natural gas burners. The advantage of the system was to reduce the amount of effluent. There was, however, an appreciable reduction in yield which according to Lee (20) would create difficulties with commercial acceptance of the system.

3.10.7 *Texture of blanched vegetables*

The texture of cooked, blanched and frozen vegetable is sometimes inferior to that of cooked fresh vegetable. In particular, cooked frozen green beans lose the characteristic crispness of the fresh product and cooked frozen carrot tends to be rather rubbery. The changes in the texture of the processed food compared to the fresh depend on the conditions of blanching, freezing, storing and cooking. As processors endeavour to speed up operations in their plant the use of short-time high-temperature blanching techniques to inactivate enzymes has become popular. These blanching techniques reduce the time the vegetables are at temperatures which soften their texture. However, other methods are available which will also improve the texture of blanched vegetables.

Steinbuch (32) investigated a blanching technique where low-temperature long-time treatment followed by short-time high-temperature treatment of vegetables produces a superior texture. He attributes the effect to the activation of pectinesterase at 70 °C during the 10 min blanch which is inactivated by the following 2 min at 98 °C. The texture of canned vegetables may be improved by the addition of calcium salts to the blanch water and this can be further improved by the addition of a small amount of citric acid to the brine. This was demonstrated with canned cauliflower by Hoogzand and Doesburg (18) where the canned product was tested with a hardness meter. Cauliflower which had a long-time low-temperature blanch measured 1.8 kg compared to a high-temperature short-time blanch which had a reading of 0.1 kg. When calcium was added to the low-temperature long-time blanch the hardness measurement rose to 4.4 kg and with acid in the brine the reading was 6.1 kg. However, with both acid and calcium in the brine the hardness value was 10.1 kg.

3.10.8 *Loss of nutrients during blanching*

The loss of some of the nutrients during blanching is inevitable. Vitamin C is both water soluble and heat labile and as such has been used by workers as an indicator vitamin when determining the effects of blanching on vegetables. Selman and Rolfe (28, 29) investigated vitamin C losses from low-temperature blanching on garden peas. Canet *et al.* (6) investigated the loss of vitamin C on blanching potatoes and found that water blanching for 2 min at 97 °C reduced the vitamin C content from 12.5 to 7.8 mg/100 g. After cooking the vitamin C content of the raw and blanched potato were 7.4 and 6.8 mg/100 g respectively.

Guerrant *et al.* (16) reported the effects of water and steam blanching in commercial machinery on the vitamin C content of garden peas, whole green beans, lima beans and spinach using dry matter or alcohol insoluble solids as the basis for calculation. Water blanching at 82.2 °C and steam blanching at 99 °C for 3 min gave similar results for vitamin C content for garden peas, whole green beans and lima beans. The results for spinach indicated large differences with the water blanched vegetable having less than half the vitamin C content of that from steam blanching. Feaster *et al.* (12), working with garden peas, found that increasing the duration of the water blanch increased the loss of vitamin C. They also found that the maturity of the peas was important because mature peas retained more vitamin C than immature ones.

Henshall (17) reviewed the loss of constituents during commercial processing operations and found variety, maturity, harvesting, handling and transport, cleaning, washing, blanching and final processing all had to be considered when determining nutrient losses in vegetables. He stated that nutrient retention was product-dependent, high-temperature short-time processes were usually desirable and a careful control of product: water ratio was necessary.

References

1. Anon., 'Waterjet cutting system ups accuracy, yields 300% productivity improvement', *Prepared Foods* **155** (7) (1986), 81.
2. Anon., 'Onions without tears', *Food Manuf.* **57** (2) (1982), 43.
3. Anon., 'Spiral vibratory blancher maximizes thermal efficiency, minimizes space', *Food Process.* **38** (2) (1977), 108.
4. Brown, G.E., Bomben, J.L., Dietrich, W.C., Hudson, J.S. and Farkas, D.F., 'A reduced effluent blanch-cooling method using a vibratory conveyor', *J. Food Sci.* **39** (1974), 696.
5. Byrne, M., 'Breakthrough in blanching', *Food Manuf.* **57** (10) (1982), 45.
6. Canet, W. and Hill, M.A., 'Comparison of several blanching methods on the texture and ascorbic acid content of frozen potatoes', *Int. J. Food Sci. Technol.* **22** (1987), 273.
7. Casimir, D.J. and Lewis, P.S., *Flame Sterilization. Specialist Courses for the Food Industry.* AIFST-CSIRO, 1972, p. 59.
8. Cowlin, R., 'TV cameras focus on quality', *Food Process.* **47** (7) (1986), 21.
9. Cumming, D.B., Stark, R., Timbers, G.E. and Cowmeadow, R., 'A new blanching system for the food industry', *J. Food Process. Preserv.* **8** (3/4) (1985), 137.
10. Decareau, R.V., 'Microwaves in food processing', *Food Technol. Aust.* **36** (2) (1984), 81.
11. Dunlop, J.J., 'New blancher needs 90% less steam and produces 90% less effluent', *Food in Canada* **41** (2) (1981), 20, 24, 26.
12. Feaster, J.F., Mudra, A.E., Ives, M. and Tompkins, M.D., 'Effect of blanching time on vitamin retention in canned peas'. *The Canner* **108** (1949), 27.
13. Gallop, R., 'Advanced "closed-loop" food/water/waste processing systems', *Aust. Inst. Food Science & Technology 1989 Conv. Proc.*, 1989, p. 155.
14. Gillespy, T.G., 'Survey of flat-sour spore infection of blanchers', The Fruit and Vegetable Canning and Quick Freezing Research Association Chipping Campden Technical Memorandum 25, 1958.
15. Graham, R.P., Huxsoll, C.C., Hart, M.R., Weaver, M.L. and Morgan, A.I. Jr, '"Dry" Caustic peeling of potatoes', *Food Technol.* **23** (2) (1969), 195.
16. Guerrant, N.B., Vavich, M.G., Fardig, O.B. and Ellenberger, H.A., 'Effect of duration and temperature of blanch on vitamin retention by certain vegetables', *Indust. Eng. Chem.* **39** (8) (1947), 1000.
17. Henshall, J.D., 'The effect of processing on the nutritive value of food', *Proc. Nutr. Soc.* **32** (1973), 17.
18. Hoogzand, C. and Doesburg, J.J., 'Effect of blanching on texture and pectin of canned cauliflower', *Food Technol.* **15** (3) (1961), 160.
19. Lazar, M.E., Lund, D.B. and Dietrich, W.C., 'A new concept in blanching', *Food Technol.* **25** (7) (1971), 684.
20. Lee, C.Y., 'New blanching techniques', *Korean J. Food Sci. Technol.* **7** (2) (1975), 100.
21. Mitchell, R.S., 'In-can blanching of green peas', *J. Food Technol.* **7** (1972), 409.
22. Ratcliffe, J.D., 'A pr ctical analysis of the in-plant peeling losses on potatoes', *Food Trade Rev.* **45** (10) (1975), 11.
23. Ralls, J.W., Maagdenberg, H.J., Yacoub, N.L., Zinnecker, M.E., Reiman, J.M., Kranath, H.D., Homnick, D.N. and Mercer, W.A., 'In plant hot-gas blanching of vegetables', *Proc. 4th National Food Processing Waste Symp, Syracuse, New York*, 1973, p. 12.

24. Ray, A., 'Steam blancher uses 50% less energy', *Food Process.* **36** (1) (1975), 64.
25. Robe, K., 'HTST pressure blancher cuts steam costs $3600/day', *Food Process.* **45** (1) (1984), 88.
26. Rumsey, T.R., Scott, E.P. and Carroad, P.A., 'Energy consumption in water blanching', *J. Food Sci.* **47** (1981), 295.
27. Schultz, R.W. and Smith, G.A., 'Additives raise peeling efficiency', *Food Engng* **40** (6) (1968), 95.
28. Selman, J.D. and Rolfe, E.J., 'Effects of water blanching on pea seeds', *J. Food Technol.* **14** (1979), 493.
29. Selman, J.D. and Rolfe, E.J., 'Effects of water blanching on pea seeds II. Changes in vitamin C content', *J. Food Technol.* **17** (1982), 219.
30. Sijbring, Ir P.H., 'The use of a brushing machine in peeling potatoes', *Food Trade Rev.* **39** (8) (1969), 41.
31. Somers, J.D., Frank, E. and Swientek, R.J., 'Optical sorting', *Food Process.* **48** (1) (1987), 78.
32. Steinbuch, E., 'Improvement of texture of frozen vegetables by stepwise blanching treatments', *J. Food Technol.* **11** (1976), 313.
33. Veldhuis, M.K. and Neubert, A.M., 'Cleaning vined canning peas by froth flotation – removal of nightshade', *Western Canner and Packer* **36** (6) (1944), 18.
34. Weaver, M.L., Huxsoll, C.C. and Ng, K.C., 'Double-dip caustic peeling of potatoes II: Studies on commercialization', *Amer. Potato Jl* **58** (1981), 355.
35. Williams, D.C., Lim, M.H., Chen, A.O., Pangborn, R.M. and Whitaker, J.R., 'Blanching of vegetables for freezing – which indicator enzymes to use', *Food Technol.* **40** (6) (1986), 130.
36. Willard, M.J., 'Infrared peeling – reduces pollution – cuts costs', *Food Technol.* **25** (2) (1971), 125.
37. Wilmont, K., 'When the chips are down', *Food Flav. Ingred., Process., Pack.* **10** (6) (1988), 21.
38. Wright, M.E., Hoehn, R.C., Coleman, J.R. and Brzozowski, J.K., 'A comparison of single use and recycled water leafy vegetable washing systems', *J. Food Sci.* **44** (2) (1979), 381.

4 Thermal processing

ALBERT HERSOM

4.1 Principles of heat processing

The basis for the preservation of food by heat is the destruction of living organisms which would lead to deterioration of the food and/or would endanger the health of the consumer. At the same time, consideration has to be given to the retention of the organoleptic and nutritional properties. Knowledge of the heat resistance of contaminating microorganisms, the chemical and physical properties of the food and the rate at which heat penetrates the product, is used to determine a heat processing regime which will satisfy these objectives.

In practice, resistance of bacterial spores is the main cause for concern, because it is in this form that bacteria are most heat tolerant. If bacterial spores in suspension are exposed to heat at a constant temperature, the surviving fractions plotted against time, in general, follow a logarithmic course (see Figure 4.1). The slope of the resulting curve is designated by the symbol D and is known as the Decimal Reduction Time. In other words it is the time of heating, in minutes, to reduce the number of the survivors to one-tenth of the original, that is, the time for the curve to span one log cycle.

The heat resistance of bacterial spores may be defined by their thermal death-time. This is the time required at a specified temperature for the inactivation of an arbitrarily chosen proportion of the population. The proportion chosen depends on the degree of assurance required against spoilage or survival of pathogens. The minimum process for low acid foods, that is, those with a pH above 4.5, is based on inactivation of the spores of *Clostridium botulinum* Types A and B, because these are the most heat-resistant pathogens. The level to which the spores should be reduced is based on the work by Esty and Meyer (1) and now confirmed by long experience in the canning industry. It is generally accepted that the 'classical' standard is equivalent to a reduction in the population by a factor of 10^{12} or $12D$.

For process evaluation, thermal destruction data at more than one temperature are required. The thermal death-time data may be presented in the form of a curve by plotting the thermal death-times on a logarithmic scale against their corresponding temperatures. Presented in this manner, the curve approximates to a straight line, the slope of which is known as the z-value.

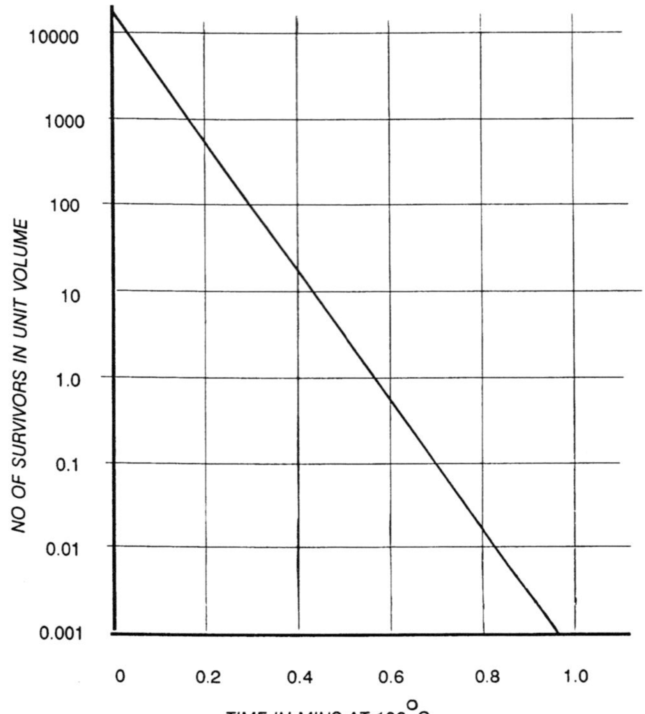

Figure 4.1 Thermal death-rate curve.

From Esty and Meyer's data, the *z*-value for *C. botulinum* is normally taken as 10 °C and the thermal death-time at 121.1 °C as 2.52 min.

To compare and quantify the sterilising effect of different heat processes, a unit of measurement has been devised. This unit is the heat process equivalent to 1 min at 121.1 °C and is designated by the symbol *F*. Using axes, as in a thermal death-time curve, straight lines corresponding to different *z*-values can be drawn through this point. All points on any one line so generated are equivalent in terms of the *F*-value. In order to define *F* completely, the value of *z* is denoted as a superscript, for example F^8, F^9, F^{10}. Because the *z*-value for *C. botulinum* and many other spores is 10 °C, $F^{10}_{121.1}$ is normally written as F_0.

4.2 Classification by pH

Because acidity is a major factor in determining the extent of the heat process required to effect commercial sterility, canned foods have been classified according to their pH. Bigelow and Cameron (2) proposed three groups as follows:

Non-acid food; pH value above 6.0
Semi-acid food: pH between 4.5 and 6.0
Acid foods: pH value below 4.5

Later, Cameron and Esty (3) suggested modifications by recognising four acidity groups to which they assigned special spoilage relationships:

Group 1. Low acid: pH 5.0 and higher.
Group 2. Medium acid: pH 4.5 to 5.0
Group 3. Acid: pH 3.7 to 4.5.
Group 4. High acid: pH 3.7 and below.

The main demarcation in the acidity classification is based on the ability of *C. botulinum* spores, regardless of type, to germinate, grow and produce toxin in foods. In spite of some reports to the contrary, the minimum pH at which this occurs appears to be 4.7 and, allowing for some margin of safety, the acidity demarcation of 4.5 is widely accepted, although the US Food and Drug Administration allows a value of 4.6.

Thus the minimum process required for low-acid foods is $F_0 = 2.52$, often referred to as a 'botulinum cook', and rounded off to $F_0 3$. For more acid foods with a pH below 4.5 (for example, most fruits), lower processes are applied because the spores of *C. botulinum* will not germinate and grow to produce toxin under these conditions. Typical pH values for canned vegetables given by Holdsworth (4) are shown in Table 4.1. Because most fall within the medium or low-acid groups there is a botulism risk and the heat process required must take cognisance of this. Because of the spoilage risk due to survival of spores which are more heat resistant than *C. botulinum*, it is generally necessary to apply higher F-values than those demanded by the

Table 4.1 pH of vegetables (from Holdsworth (4)).

Asparagus	5.5–6.0	Celery	5.3–6.0
Beans in tomato sauce	5.3–5.9	Corn	6.1–6.3
Beans: broad	6.0	Eggplant	5.6
butter	6.0	Mushrooms	6.2
dwarf	5.2–5.6	Olives	6.9
stringless	5.2–5.6	Parsnip	5.1
runner	5.2–5.6	Peas	6.0–6.2
lima	6.2	Potato	5.4–5.8
red kidney	5.9	Pumpkin	5.3
soya green	6.6	Sauerkraut	3.5
Beetroot	5.2–5.5	Spinach	5.4–6.0
Carrot	5.1–5.3	Swede	4.7–5.1
Cauliflower	5.7–6.1	Turnip	5.5

'botulinum cook'. Furthermore, some vegetable packs require increased processes in order to reach an acceptable texture.

4.3 Heat penetration

In the normal canning process, the unsterile food is packed and sealed in the container before the heat process is applied. To evaluate this, it is necessary to take into account the contribution made while the contents are heating-up and cooling-down. Thus, the rate of heat penetration into the product has to be measured and this is normally achieved by placing the hot junction of a thermocouple at the slowest heating point in the container and plotting the indicated temperature against time. The positioning of the hot junction is dependent on the physical properties of the product, and whether the pack is subjected to mechanical agitation during the heat treatment. Solid foods, or solidly packed foods without added liquor, heat by conduction and the slowest heating point is normally taken as just above the geometric centre of the can. Packs containing adequate liquid of moderate or low viscosity heat largely by convection, and the slowest heating point is taken to be on the central axis at a point between the centre and the bottom of the can. Some packs, such as those with a high solids to liquids ratio, or those with large solids in a thin liquor, heat both by conduction and convection. In such cases, the positioning of the probe has to be decided from the nature of the product, but is likely to be at the centre of any large piece. In flexible pouches the probe is generally located at the centre of the pouch, or within any large solid piece as appropriate to the product. It has been argued that packs such as whole potatoes in brine, may not require a 'botulinum cook' at the centre of the potatoes because infection is low or non-existent at this point. The possibility of damage to the tubers, which would allow contamination from bacterial spores but not free access to convective currents, makes this an unwise procedure.

Ball (5) observed that foods in which rapid heat penetration is achieved, usually were of better quality when sterilised at temperatures above 121 °C than when sterilised at lower temperatures. To explain this, it is necessary to examine the effects of different time and temperature treatments on the organoleptic and nutritive properties of the food and to compare these with the thermal death-time data for bacterial spores. One method of making this comparison is to superimpose time/temperature data for the loss of nutritional and quality indicators on to the graph used to plot the thermal death-time curve for bacterial spores. While the slope of the curve (z) for spore destruction is normally about 10 °C, that for nutritional indicators such as thiamin destruction (6, 7), degradation of chlorophyll (8, 9) and the oxidation of ascorbic acid (10) is nearer 33 °C.

The change in sensory quality during thermal processing follows a similar pattern, although widely different z-values have been reported. Mansfield (11) proposed an equation for depicting cooking effects which is analogous to that for sterilisation with the unit of cook being taken as 1 min at 100 °C. This is written as:

$$C = t \cdot \log_{10}^{-1} \frac{T - 100}{z_c}$$

where C is the unit of cook; T is temperature in °C; z_c is the slope of the cooking curve analogous to the thermal death-time curve defined earlier; and, t is the time in minutes. Mansfield used a value of 59.7 °F or 33 °C for the z_c-value. Hypothetical cooking curves at different temperatures with z-values taken as 33 °C are shown in Figure 4.2. Because of the differing slopes between these

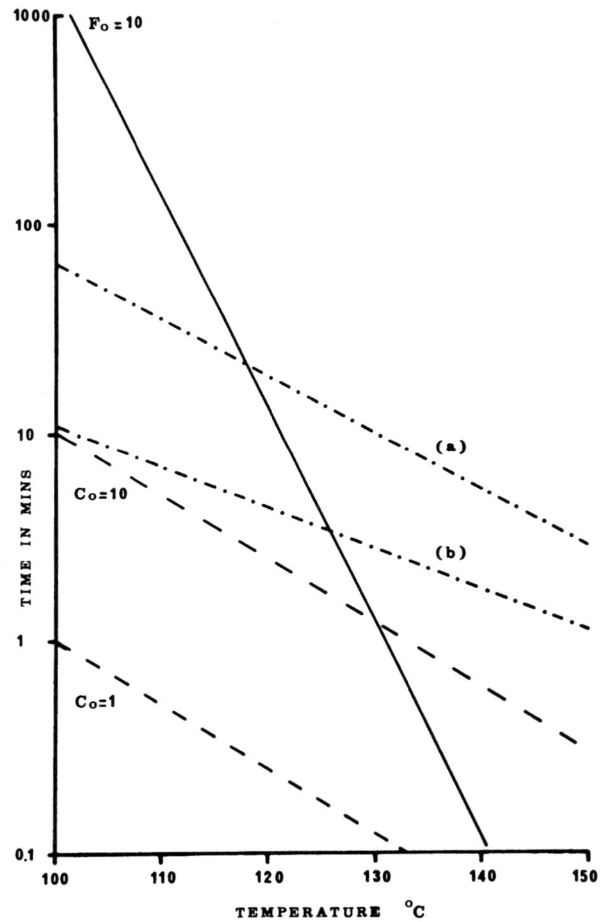

Figure 4.2 Comparitive cooking and sterilisation curves. (a) Lipase—P_s fluorescence (90% inactivation); (b) Peroxidase—irreversible inactivation.

and those for inactivation of bacterial spores, it is to be expected that foods of better quality would be obtained if the sterilisation process was to be carried out by means of a high-temperature short-time (HTST) process, rather than by the traditional canning process in which the temperature is usually limited to about 121 °C. For some products heat penetration is aided by agitation, which enables higher processing temperatures to be used; for others, the heating and cooling of the product is carried out on-stream using heat exchangers to speed up penetration, after which it is filled into sterile containers which are sealed under aseptic conditions.

Unfortunately, the anticipated benefits of the HTST process may not be fully realised in all cases due to the survival of enzymes. For example, it has been shown that canned peas processed to F-values in the range of 5–12 at 127 °C or 3–7 at 121 °C, develop viny off-flavours on storage (Guyer and Holmquist, 12) Processing to higher F-values at these temperatures, or processing to similar F-values at 115 °C, did not lead to off-flavour development. This effect was shown to be due to peroxidase activity which was regenerated on storage in those samples processed at the HTST options or at the lower F-values. This effect may be attributed to the high D- and z-values of peroxidase. Adams (13) reported D_{120} values of 3.78 min and z-values of 34.5 °C for the irreversible inactivation of horseradish peroxidase in pea purée.

4.4 Containers

The thermally processed food must be protected by a suitable container the main purpose of which is to prevent the ingress of microorganisms and oxygen. The two most widely used for processed vegetables are the tinplate can and the glass container.

4.4.1 *Cans*

The three-piece can remains the most economical, reliable and widely accepted form of tinplate container and constitutes the bulk of those produced. Modern tinplate is made by electrolytically depositing tin on to thin steel baseplate which may vary in thickness from 0.166 mm (double reduced) to 0.25 mm. Baseplate specification is critical. Low phosphorus and trace metal content is necessary if good corrosion resistance is required, as is the case for some fruits. Tin coating weight is expressed in grams per metre squared (g/m^2) and varies from 1.4 to 11.2, the actual amount depending upon the condition applying, and whether a protective lacquer is applied. The tin coating weight may need to be different for the internal and external surfaces of the can. To satisfy this requirement, differentially coated plate is available and widely used.

The three-piece can is made from a cylindrical body and two ends. To form the body, sheet cut from the coil is formed into blanks which are notched at the corners, so that when the side seam is formed the ends of the seam comprise only two thicknesses of metal. The side seam is formed by hooking the edges and locking these together to form the cylinder. Solder flowed into the seam completes the seal, after which the edges are flanged ready to receive the can ends. The can ends are stamped from tinplate and sealing compound is deposited in the curl at the edge. Once the end is positioned on the flange of the can body it can be formed into a seam by a two stage operation consisting of (a) rolling the curl of the end under the flange of the body and (b) ironing these together in a precise manner. The sealing compound serves to fill minute spaces within the rolled seam to form a hermetic closure. Welded side seams, which use a simple overlap joint, are now widely used and these have advantages over soldered cans. Double seaming of the ends is more reliable, the side seam is more durable, near all-round lithography is possible and tin-free steel may be used in the construction.

Two-piece cans, made either from tinplate or aluminium, are popular for some applications, for example, in the beverage industry. These are made by drawing a can complete with base from a single disc of metal. Developments in the methods of producing drawn cans enable full height cans to be made by this process. Further information on the methods of manufacture are given by Ings (14).

4.4.2 *Glass containers*

Glass containers are suitable for a wide variety of heat-processed foods. When provided with a suitable closure, the jar or bottle provides an inert, hermetic, durable and transparent package which is well suited for processed vegetables. Glass containers for food are available with different surface treatments designed to protect them from abrasion and to increase lubricity which makes high-speed handling easier.

Containers used for sterilised low-acid products such as vegetables are sealed by vacuum hermetic closures, the three types in general use being the pry-off (side seal), the lug-type (twist-off), and the PT (press-on twist-off). These caps are applied and sealed in steam-flow cappers, and retained in place during the sterilisation process by the vacuum so formed and an applied overpressure.

4.4.3 *Flexible pouches*

A variety of sterilised low-acid foods in retort pouches have been produced and widely accepted in Japan for over 20 years. In Europe and the USA they have not gained such wide acceptance in spite of their technical suitability. The thin cross-section allows shorter processing times than those required for cans or glass containers of the same capacity and, for some products, this results in improved quality. This effect is most marked for the larger sizes as used for catering packs which seems to be the most promising application.

The structure of the pouch in general use is a three-ply laminate of polyester/aluminium foil/polypropylene in which the polypropylene forms the inner seal and the aluminium an oxygen and light barrier. Pouches of this make-up withstand sterilisation temperatures of up to 130 °C which enables an HTST process to be applied thus taking advantage of the thin cross-section.

Many commodity foods such as corn, green beans, sliced or diced carrots, as well as speciality packs, such as potato salad, are candidates for this method of packaging.

4.4.4 *Trays*

Semi-rigid two-piece trays made from aluminium and heat sealed with an aluminium foil cover have been used in Europe and Japan for thermally processed foods. The preferred construction for the tray and lid is aluminium laminated to polypropylene. Containers of this construction are heat sealable and may be processed in superheated water with an overpressure, or in a steam–air mixture. Some of the faster heat transfer characteristics of the pouch are present although to a lesser extent. Compared with conventional cans, material costs are higher and line speeds slower.

In the USA the half-size steam table tray has been used for institutional (catering) packs. The tray is made from chrome steel plate pre-coated with a vinyl lacquer. The lid of similar composition is double seamed like a standard three piece can. Similar advantages and disadvantages to the aluminium foil tray apply.

4.4.5 *Thermally intolerant containers*

For use with aseptic systems, a number of containers may be considered including thermoformed cups and trays, cartons made from paperboard/foil/thermoplastic complexes and bulk packs such as bag-in-box combinations. A characteristic of these containers is that they may be sterilised by chemical or physical means independently of the food. This is discussed further in the section on aseptic processing.

For further information on containers see Chapter 8 on Packaging.

4.5 Outline of canning operations

The canning of vegetables may be considered under the following operations:

1. Preparation of the food
2. Filling
3. Exhausting the container

4. Sealing
5. Thermal processing/sterilisation
6. Cooling the container and contents

4.5.1 *Preparation of the food*

A number of processes are involved such as washing, grading, peeling and blanching.

4.5.1.1 *Washing.* This usually involves agitating or tumbling the vegetables on moving belts or revolving screens while they are immersed in water or subjected to water sprays. Washing by means of high-pressure water sprays is the most satisfactory method. Sprays are effective only if the water reaches all parts of the product. This can be achieved if the sprays are directed from above and below a travelling woven wire-cloth conveyor, or by designing the washer so that the product is induced to roll over during the spraying process.

4.5.1.2 *Grading.* Many products are size graded by passing them over screens with holes of different diameters. Some vegetables are passed over a series of parallel bars with varying distances between the bars. In either case the screens may revolve or vibrate.

4.5.1.3 *Peeling.* This is an important preliminary operation since, in conjunction with washing, it removes surface soiling and associated microbial contamination. Methods used include steam, mechanical, flame, abrasive and lye peeling.

4.5.1.4 *Blanching.* Blanching of vegetables is carried out prior to the canning process for the following reasons: (1) to remove respiratory gases which would reduce the ultimate vacuum in the can if they were released during processing; (2) to inhibit enzymic reactions which might occur prior to the heat-processing stage; (3) to promote shrinkage of the product thus permitting an adequate fill of the container; (4) to hydrate certain dry products which are not fully rehydrated prior to the blanching operation; and (5) to pre-heat the product in order to assist the vacuum formation in the can.

Different methods of blanching have been proposed, the most common being to immerse the vegetables in hot water at 85–95 °C. The rotary immersion blancher, in which the product is conveyed by a screw on the inside of a rotating reel, the bottom part of which is immersed in the hot water, is widely used for this purpose. Chain conveyors are also used for transporting products through the hot water. In hydraulic-type pipe blanchers, the product is pumped with hot water through a continuous length of pipe; steam jets along its length being used to heat the water and assist the flow.

Steam blanching is also used; the continuous steam blanchers being mechanically more complex, because the product has to be moved through a steam chamber. Ray (15) has described a steam blancher with hydrostatic seals which is claimed to be more energy efficient than conventional steam blanchers.

Because hot-water blanching causes loss of some nutrients and gives rise to large volumes of waste water, other methods have been proposed. Hot-gas blanching has been described by Robe (16) and microwave blanching has been proposed. Ralls (17) compared the volume of water required for blanching vegetables by microwave, hot air, steam, and hot water. He reported that hot-air blanching required least water, followed by microwave, steam, and water methods. The IQB (individual quick blanch) method has been described by Lazar *et al.* (18), the principle of which is to subject each particle for a relatively short period to an atmosphere of steam, and then to accumulate the product in a deep bed without further application of heat until equilibration of temperature takes place. Results from trials indicate that IQB and hot-gas blanching give products of similar quality to those obtained from steam or hot-water blanching but with reduced nutrient loss and up to 99% less effluent.

4.5.2 *Filling*

Filling into glass or metal containers is accomplished mechanically or by hand. Apart from the economic aspects careful control of the filling weight is important for technical reasons. The volume of the headspace may affect the efficiency of the exhausting procedure and, if an agitating process is used, the rate of heat penetration into the container. The ratio of solid to liquid material also influences the rate of heat penetration thus affecting the extent of the heat-processing treatment.

4.5.3 *Exhausting the container*

Removal of air from the container before it is closed is an important operation. Apart from minimising strain on the can seams or pouch seals during processing, removal of oxygen helps to conserve quality and reduce internal corrosion. The creation of a vacuum after cooling ensures that the ends remain flat or slightly concave throughout moderate changes of storage temperature or barometric pressure, thus providing assurance to the consumer of the integrity of the container.

Vacuum in the container may be achieved by means of a heat-exhaust, by mechanical means, or by steam injected into the headspace just prior to sealing. The heat-exhaust method depends on the product. If the product can be filled hot this suffices, but otherwise it may be necessary to pass the filled containers through a steam tunnel prior to sealing. Mechanical methods include sealing cans in a vacuum closing machine, sometimes preceded by vacuum syruping. In the headspace injection method, steam is flushed through ports around the seaming head of can closing machines in order to sweep out air from the headspace. Steam flow is generally used during sealing of glass containers. The method is effective if a covering syrup or brine is present, but may not be suitable for uncovered solids containing occluded air. Pouches may be sealed in a partial vacuum, although the use of a pressure plate which squeezes the pouch during sealing is often adequate.

4.5.4 *Sealing*

Cans are closed by the formation of a double seam of the same construction as that used on the can maker's end. These may, or may not, be vacuum sealed. Glass containers for processed products are generally vacuum sealed. Both can and glass containers may be closed at high speed. The sealing of flexible pouches, which relies on the fusion of two thermoplastic materials, is a slow operation and the high speeds obtainable for cans and glass containers are not yet possible.

4.5.5 *Thermal processing (sterilisation)*

The sealed containers have to be processed in order to effect sterilisation of the contents. This is usually achieved by heating for an accurately predetermined time and temperature in an atmosphere of saturated steam, in heated water, or, occasionally, in a steam–air mixture. The heating medium used is largely determined by the type of package, those requiring excess overpressure, such as glass and pouch packs, being processed in water with air overpressure, or in a steam–air mixture.

The most widely used heating medium is saturated steam, which transfers heat to the product by virtue of a high heat transfer coefficient and high latent heat of condensation. Absence of air is vitally important and is normally ensured by a procedure known as venting. This is aimed at replacing all air in the retort by steam before it is brought to operating temperature. To ensure a sound venting operation the design of retorts – particularly the positioning and sizing of the vents is important. Where air is deliberately introduced as in steam–air mixtures, the mixture must be uniform throughout the process. Special steam–air mixer valves and fans in the retort have been used for this purpose.

Processing may be carried out in batch or continuous retorts, the latter being equipped with specially constructed ports or hydrostatic seals, through which the containers enter and leave the pressure zone. In both batch and continuous retorts the containers may be stationary throughout the process, or they may be moved or agitated in order to speed up the penetration of heat into and out of the container.

4.5.5.1 *Batch retorts.* Saturated steam retorts are specially designed pressure

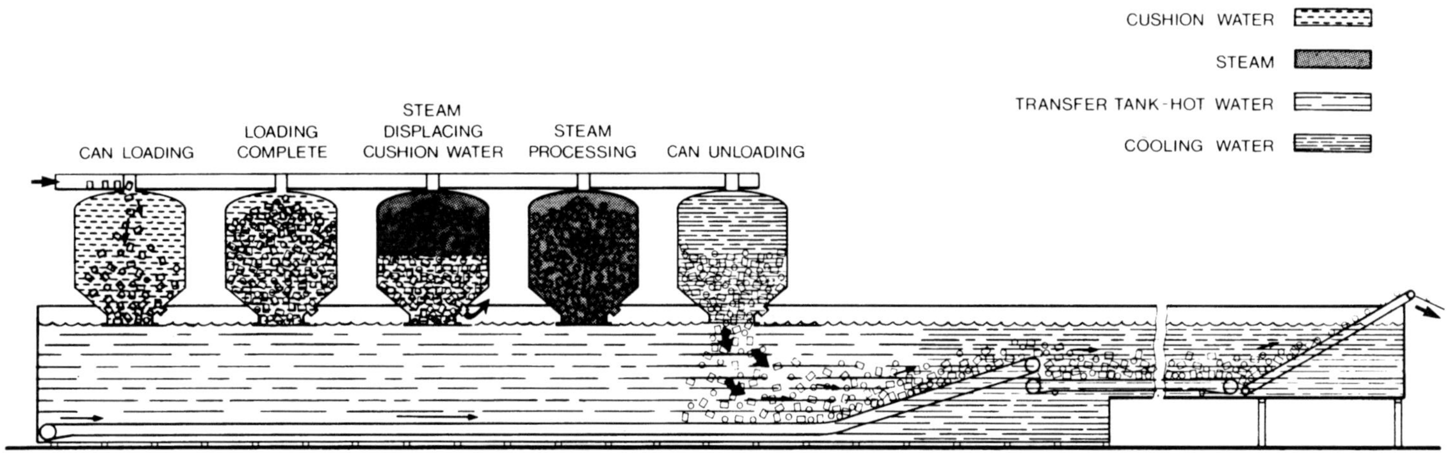

Figure 4.3 Crateless retort (courtesy FMC Corporation).

vessels into which the cans are loaded, usually stacked or jumbled into crates or cars. They may be vertically or horizontally disposed. Because venting and steam distribution are of paramount importance their design is critical and details of design and operational procedure may be found in the NFPA Bulletin 26-L (19). Bulletin 30-L (20) gives details of overpressure retorts used for glass processing.

A derivative from the vertical batch retort is the crateless retort shown in Figure 4.3. By means of a conveyor, cans are loaded through a top entry into a retort which is filled with water. After the required number of cans have been loaded, a gate on the conveyor closes and the next retort is filled. The water, which may be pre-heated, acts as a cushion for the cans falling through the top opening. After the retort is closed steam is admitted through a spreader at the top which forces out the water. This may be collected, reheated and used in another retort. At the end of the process the cans are unloaded through the bottom into the cushion canal, which is equipped with a conveyor to move them to the cooling canal. The cans may be pressure cooled in the retort if desired.

Batch retorts in which the cans may be agitated include the Rotamat and the FMC Orbitort. The actions of the two systems are different, the Rotamat providing an end-over-end agitation while the Orbitort rotates cans about their vertical axes. The heat penetration rate is improved in many products by agitation, thus reducing processing times and enabling higher temperatures to be used.

4.5.5.2 *Continuous retorts.* These may be of the rotary type in which cans enter and leave the pressure zone through self-sealing rotary valves, or of the hydrostatic type in which the mechanical valves are replaced by columns of water which balance the internal pressure.

The most widely used rotary type is the FMC Sterilmatic cooker-cooler, in which cans are carried in compartments at the circumference of a rotating reel. The reel pushes the cans through a spiral track attached to the inside of the shell, the lead of the spiral advancing the cans through the chamber. A three-phase action is applied: (i) free rotation at the lower portion of the shell where the cans roll in the spiral; (ii) fixed reel travel in the upper portion where the cans sit in the pockets and are carried about a central axis, and (iii) a sliding action in the transitional stage between the two (see Figure 4.4). At the end of the steam chamber the cans are transferred to the cooling shell, or shells, where they are normally cooled under pressure.

In the Hydrolock continuous cooker-cooler, agitation is achieved by cans rolling on a fixed perforated stainless-steel plate as they are pushed through the equipment by carriers. The containers enter via a rotary paddle wheel with spring-loaded sealing bars. The water, which exits through the chain openings, flows over a weir into a return line. The system may be used with saturated steam as the heating medium, although, when an overriding pressure is

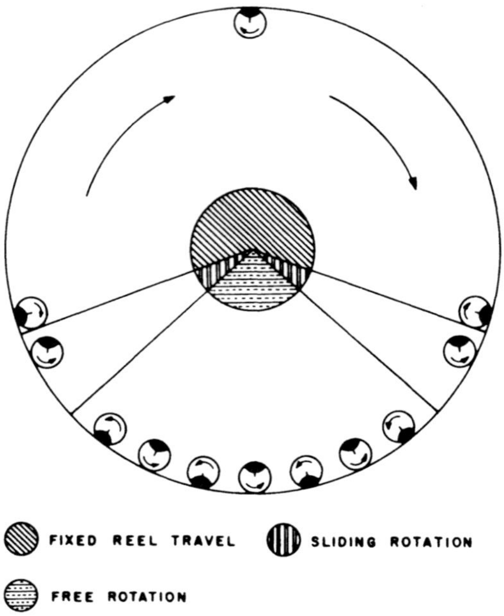

Figure 4.4 Can motion in Sterilmatic cooker-cooler.

required as with glass containers or retortable pouches or trays, an air–steam mixture may be provided and maintained by one or more turbo fans.

Hydrostatic cooker-coolers comprise 'bring-up' and 'come-down' legs which balance the internal pressure of the steam chamber. The containers, which are conveyed through the machine by means of carriers connected to heavy duty chains, are subjected to gradual pressure and temperature changes as they enter and leave the sterilising zone. Additionally, because only gentle rotation is applied, hydrostatic cookers are particularly gentle in treatment of product and container alike. The containers leave the 'come-down' leg partially cooled, the cooling operation being completed by water sprays. Because water in the exit leg may be at a temperature suitable for the growth of thermophiles, care must be exercised in the design and use of hydrostatic cooker-coolers to avoid growth of these organisms, or contamination may result as the containers are prone to microleakage during the early stages of cooling.

An alternative non-agitating steriliser is the 'Storklave' which processes containers stacked on pallets. Filled pallets are conveyed by a rack and pinion movement into a pressure lock, in which steam is introduced to purge air and to pressurise the chamber – this in order to equilibrate the pressure in the lock to that of the retort. In the retort chamber the pallets are engaged by slowly turning screws which raise the pallets up and through the retort. The pallets leave at the top through a 'come-down' lock, which is then vented to an appropriate lower pressure for discharge into the cooling tower. The containers are cooled by water entering the top of the tower and cascading to the bottom. Since cooling is accomplished in about half the time required for sterilising, there is spare capacity in the cooling tower which can be used for can drying. This is recommended in order to reduce leaker spoilage.

The FMC Universal Steriliser is designed for the processing and cooling of products in virtually any container, regardless of shape, size or construction. It consists of a horizontally aligned pressure vessel, with a lock at either end for entry and exit of cars which house the containers on stainless-steel trays. Each car is automatically transferred to the infeed lock where the containers are pre-heated with hot water and the external pressure is adjusted to that of the main chamber. The car moves on a track through the pressure vessel where heated water flows over the containers. A novel feature is the manner in which the water enters at the top of the pressure vessel directly into the car and then flows either horizontally or vertically over the containers. The water does not fill the pressure vessel itself. The discharge lock provides pressure cooling for the containers, although final cooling takes place outside the vessel. The ability to regulate the overpressure in the chamber makes the system suitable for the processing of flexible pouches, trays or glass containers.

4.5.5.3 *Flame sterilisation.* This method, which relies on the direct application of gas flames to the outside of the container which is rotating during the process, has been used as a method of heat processing canned products such as mushrooms. A pre-heating section, which uses steam at atmospheric pressure to bring the contents to a uniform minimum temperature, is an essential step in this process. Thin liquids, or solids packed in thin liquids, are suitable products for flame sterilisation. With these products heat is rapidly transferred to the contents of the can and HTST processes may be applied. The early Steriflamme equipment is described by Beauvais (21) and more modern equipment such as the CSIRO flame steriliser by Casimir (22). Leonard *et al.* (23–25) discuss the principles and theory of flame sterilisation, and more recently Richardson (26) has reviewed the process. Different sterilisation systems are described in detail by Lopez (27).

4.5.5.4 *Microwave sterilisation.* The use of microwaves for sterilising food is attractive and was suggested by Benjamin and Ecklund (28) in the early days of microwave heating. Continuous sterilisation of plastic pouches of food in a microwave tunnel was investigated by Kenyon *et al.* (29) and Ayoub *et al.* (30) in the US Army Natick laboratories. These and other studies have shown that it is difficult to maintain reliable sterility due to non-uniform temperature distribution. Workers at SIK and Alfa-Laval AB believe some of these irregularities, for example, corner and edge overheating, can be overcome by submerging the food packages in water during the microwave treatment (31). This method has been embodied in the Alfastar Multitherm process.

4.6 Aseptic processing and packaging

Aseptic packaging of foods enables products, sterilised in bulk or on-stream, to be filled and sealed in sterile containers under aseptic conditions. There are two main reasons for its use: (i) to permit the use of containers which are unsuitable for in-package sterilisation and (ii) to take advantage of high-temperature short-time (HTST) processes which are thermally efficient and generally give rise to food which is superior in quality to that produced by processing at lower temperatures for longer times.

It should be noted that the term HTST is used in the dairy industry to describe pasteurisation processes with time–temperature profiles in the order of 72 °C for 15 s. On-line sterilisation of dairy products, carried out within the temperature range of 135–150 °C, are called UHT processes. In this text the term HTST is used to describe a process which brings about sterilisation by heating a product for times ranging from a few seconds to 6 min, as defined by Ball and Olson (32).

4.6.1 *Sterilisation of the product*

Apart from the in-can processes such as the use of agitating cooker-coolers, there are two methods of applying HTST techniques:

1. By heating the food to a high temperature in bulk or on-stream, after which it may, or may not, be partially cooled before filling. The hot product is filled and sealed in the can at a sufficiently high temperature to sterilise the container and lid. It is then cooled in the can in the normal manner. For products with a pH above 4.5 this involves filling at temperatures above 100 °C under pressure as in the Flash 18 process (33).
2. By an aseptic process. This involves heating the food to the operating temperature, normally by pumping through a suitable heat exchanger, and holding it in bulk or on-stream, to effect sterilisation, after which it is cooled, filled and sealed in sterile containers under aseptic conditions.

Aseptic processing has been restricted to liquids, or liquids containing small particles, due to the technical difficulties associated with the processing and handling of foods containing larger particulates. Some of these difficulties are:

1. Heat penetration into the particle is a limiting factor, necessitating a long holding section which is detrimental to quality.
2. Flow rates of the particles are difficult to evaluate and there may be considerable differences in transit times between individual pieces.
3. Processed solids are often fragile and difficult to handle without breakage.

4. There is a danger of overprocessing the sauce in which the pieces are suspended.

4.6.1.1 *Processing systems.* Heat exchangers used for HTST or UHT processing may be classified as follows:

Indirect heat exchangers. The product and the heat exchange fluid are separated by the heating surface. There are three types: (i) tubular heat exchangers; (ii) plate heat exchangers and (iii) scraped surface heat exchangers.

Direct heat exchangers. Steam is condensed in the product for heating and vapour is removed for cooling. There are two types of heater: (i) steam injection (steam into product) and (ii) steam infusion (product into steam).

For products containing particles of a size which would make the above systems unsuitable the Jupiter or Ohmic Heater systems from APV may be appropriate. The Stork Steripart system, offered for liquid foods containing particulates, introduces a temporary and selective holding section into the conventional heat exchanger systems. Different processing systems are offered by manufacturers and may be classified on the basis of the heat exchange equipment used. A summary of equipment is given in Table 4.2 and is detailed by Hersom (34). Many of these systems have been developed for the UHT treatment of milk and other dairy products, although most are suitable for the processing of other fluid products such as vegetable juices and purées. Many low-acid vegetable juices are difficult to process because colloidal matter in suspension is precipitated during the heat treatment. However HTST processing equipment equipped with an aseptic homogeniser down-stream reduces the settling of solids (35). Stephens *et al.* (36) studied the processing of carrot juice and suggested a method of stabilisation based on an acetic acid blanch of the raw carrots before juice extraction and canning.

Although tomato juice is suitable for hot filling it is first 'flash' sterilised in a tube or plate heat exchanger at 121 °C to destroy the flat-souring organism *Bacillus coagulans.* It is cooled to a temperature just below 100°C before filling at not less than 90 °C. If it is to be filled into containers which will not withstand this temperature it must be cooled to 35 °C or less and filled aseptically.

Canned diced, sliced or whole vegetables packed in brine or syrup are not processed aseptically: indeed, there would be little point because good heat transfer is readily obtained in conventional retorts enabling reasonably high process temperatures to be used. However, if there is a requirement for a package, for example, one based on thermoplastic, or on paper-board and thermoplastic, which will not withstand the process temperature then aseptic packaging would need to be considered, although, as pointed out earlier there are technical difficulties in this approach.

Table 4.2 Aseptic processing plant.

Type	Manufacturer	Country of origin
(a) *Indirect heating*		
Tubular	APV (Paracel)	UK
	Cherry-Burrell (Spiratherm)	USA
	Nuova Frau (Sterlux)	Italy
	Stork (Sterideal)	Netherlands
	Ahlborn	West Germany
	Alfa-Laval (Stertherm)	Sweden
Plate	APV (Ultramatic)	UK
	Sordi	Italy
(b) *Direct heating*		
Steam-into-product (Injection)	Alfa-Laval (VTIS)	Sweden
	APV (Uperiser)	UK
	Cherry-Burrell	USA
	Rossi & Catelli	Italy
	Stork (Steritwin)	Netherlands
Product-into-steam (Infusion)	Crepaco (Ultratherm)	USA
	Laguilharre	France
	Nuova Frau Dasi	Italy/USA
	Pasilac (Palarisator)	Denmark
(c) *Other types*		
Electric resistance heating	Elecster	Finland
	APV Ohmic Heater	UK
Friction	Atad	France
Particulate systems	Jupiter	UK
	Stork Steripart	Netherlands

Although a considerable amount of time is being devoted to development of the processing and packaging of products containing large particulates (greater than 8 mm cross-section) there are currently only three processing systems offered by manufacturers namely, Jupiter, the Ohmic Heater and Steripart.

Jupiter. This system is described by Hersom and Shore (37). The solids are heated and cooled using a sequential batch method while the liquor or sauce is sterilised continuously. In order to provide good heat transfer to the solids, they are tumbled in a double cone-shaped vessel which rotates about a central axis, into which steam is admitted under pressure. Heat is transferred to the solids from the steam and the condensate which is formed. The latter lubricates the pieces and helps to prevent abrasion and damage. If the condensate is insufficient for this purpose, pre-heated liquor obtained from a previous batch, or superheated water, may be pumped into the vessel during the heating cycle in order to supplement the condensate. Because the injected liquid may be at a temperature in excess of that of the solids, it also helps to speed up the heating of the solids. The vessel has a steam jacket which is maintained at the same temperature as that in the double cone. This guarantees that there are no cold

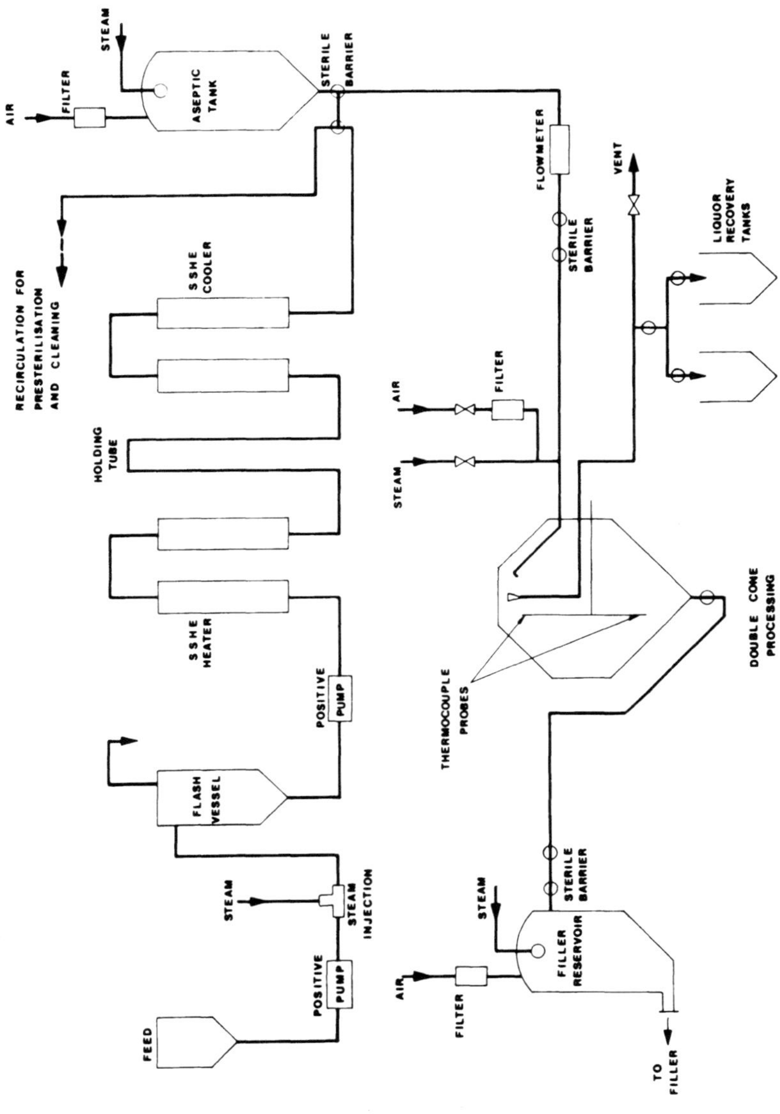

Figure 4.5 Jupiter system.

spots in the vessel; it also ensures temperature uniformity across the wall, and this, in conjuction with the washing action of the condensate prevents burn-on. The end of the process is indicated by a lethality indicator which uses data generated from a temperature probe in the product The steam in the double cone is then replaced by sterile air and the steam supply to the jacket is replaced by cooling water. The vessel continues to rotate, cooling its contents of solids and cooking liquor. At the end of the cooling cycle, rotation of the vessel is stopped and the liquor removed via a drain valve. This liquor may be retained for use as injection liquor in a subsequent batch, used as a component of the sauce, directed to the filler as a top-up diluent or discarded according to the product's needs.

Sauce is generated continuously from a standard HTST (UHT) plant and accumulated in an aseptic reservoir. As required, it is metered into the double cone containing the cooled sterile solids which then acts as a mixer. The cooled solids, now protected by the sauce against damage, are removed from the double cone simply by opening a valve and allowing the overpressure to force out the contents. The configuration of the system may be seen by referring to Figure 4.5.

Ohmic Heater. If an electric current flows in a substance of suitable conductivity heat is generated within the product. The Ohmic Heater, for which APV holds the licence rights, exploits this principle in order to heat products in a continuous manner to temperatures suitable for HTST processing. The extent of the heating depends upon the electrical conductivity and the residence time in the heater. Insulators, such as fats, oils, sugar syrups and non-metallic solids such as bone and cellulose cannot be directly heated by this method but many food products conduct electricity sufficiently well to be suitable for ohmic heating. Particles in suspension heat at the same rate as the carrier if the conductivity of each of the two components is the same.

The main element of the system is a vertical column along which are disposed the electrode assemblies. These consist of coated titanium electrodes in a PTFE housing. The electrodes, which are hollow to allow a small flow of cooling air, are connected to each phase of the secondary coil of a transformer. Interconnecting tube sections, made of stainless steel with electrically insulating plastic linings, are bolted to the electrode housings by means of flanges. The number of electrodes depends on the capacity required, but inlet and discharge electrodes are always earthed for safety (see Figure 4.6).

The required voltage is typically 3.3 kV per phase which is provided from a three-phase isolating transformer. Power output is controlled by a thyristor on the primary side and this is used for regulation of process temperature.

The system has been proved on a 45 kW unit on which different products, including fruits and vegetables, have been processed satisfactorily. Retention of particulate integrity is excellent and, in a properly managed unit, sterilisation appears to be satisfactory. Cooling, and in some cases,

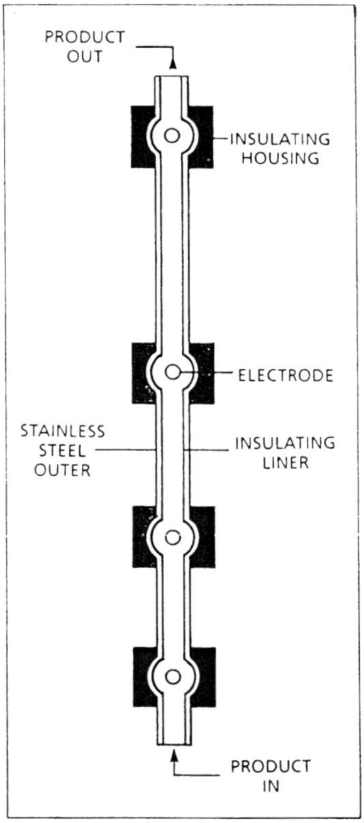

Figure 4.6 Principle of ohmic heating.

sedimentation of the particles, may be a problem on large installations and, as in all systems, indeterminate times of particles in the holding tube remains a problem. The first commercial application is imminent and initially is being confined to bulk packs for caterers and processors (38).

Steripart. A technique developed by Stork, termed 'single-flow fraction specific thermal processing (single-flow FSTP)' aimed at increasing the retention time of the particulates in an otherwise conventional heat exchanger system, has been embodied in the Steripart processing equipment. Particles are sieved from the carrier liquid in the hot hold section of the plant, increasing the time they are held in the hot liquid. Two types of sieves are proposed; (i) a Rota-Hold device to separate two fractions only and (ii) a Spiral-Hold for separating several different size fractions.

The Rota-Hold consists of a cylindrical vessel provided with an infeed and a discharge port (see Figure 4.7). Fork blades mounted on a central axis rotate slowly at a speed which may be adjusted. The solids which do not pass the

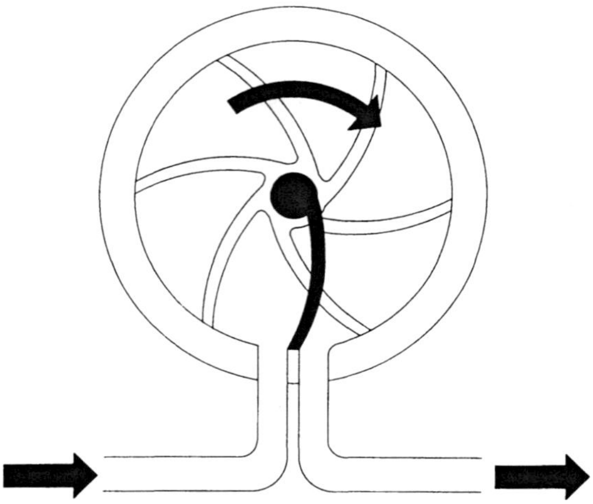

Figure 4.7 Stork Rota-Hold (courtesy Stork Amsterdam BV).

blades are conveyed around the cylinder until they are released at the discharge port. The liquid can flow freely between the pins reaching the outlet port in a much shorter time.

The Spiral-Hold consists of a tube-shaped vessel, positioned vertically, with the infeed at the top and the discharge at the bottom (see Figure 4.8). A central shaft fitted with spokes mounted spirally turns slowly at an adjustable speed. The inside wall of the tube is also provided with spokes which are mounted spirally but in a contrary pattern to those on the shaft. These extend as far as the shaft and there is also minimum clearance between them and the rotating spokes. Moving pockets are formed between the spokes and the wall, carrying solids at a predetermined rate through the system. By varying the thickness of the spokes, either in steps or gradually into the flow, the separating characteristics can be varied for different particulate sizes imparting shorter residence times with decreasing sizes. The processing equipment embodying the above devices is known as the Steripart system and includes concentric double-tube heat exchangers with heat regeneration, pumps, pressurised buffer tanks and full PLC control with instrumentation.

4.6.1.2 *Packaging systems.* Systems are available for the aseptic filling and handling of different packages, ranging from the traditional tinplate can and glass bottle to the non-rigid and semi-rigid containers based on thermoplastics or combinations of thermoplastic with paper-board and metal. In addition to those for retail packs, bulk packaging systems are used for conserving raw material or intermediates intended for reprocessing or for use in catering establishments.

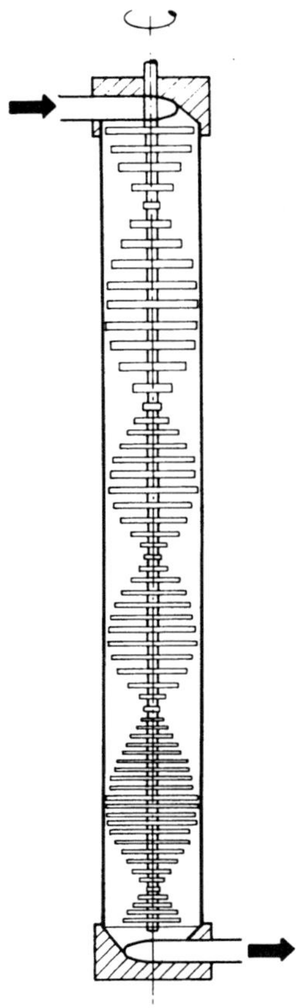

Figure 4.8 Stork Spiral-Hold (courtesy Stork Amsterdam BV).

For convenience the systems may be classified according to the following package types:

1. Cans and bottles
2. Sachets and pouches
3. Cups
4. Cartons
5. Bulk packages.

Cans and bottles. The different systems are summarised in Table 4.3. The Dole process, developed in the 1940s, uses superheated steam for sterilising the cans and providing sterility in the filling chamber. The standard filler in the Dole equipment is a simple slit through which the product flows continuously into the cans. The level is adjusted and controlled by the empty can conveyor line speed. One of the early applications was for split-pea soup (39, 40) but any

Table 4.3 Aseptic filling systems – cans and bottles.

Manufacturer	Form	Material	Sterilising and filling method
Dole	Cans	Tinplate, aluminium	Super-heated steam used for sterilising cans and in the filling chamber.
Dole	Cans	Composite of laminated foil/paper	Hot air used for sterilising containers and in the filling chamber.
Rommelag	Bottles	Granulated thermoplastic	Hot parison blown and filled immediately.
Serac	Bottles	Glass or preformed plastic	Chemical sterilisation of containers. Filtered air in the filling chamber.
Holopack	Bottles	Granulated thermoplastic	Extruded tube is cut to length and the parison at the top of the mould is ballooned with sterile air. The mould is transferred to a filtered air chamber for blowing, filling and sealing.
Remy	Bottles	Granulated thermoplastic	Hot parison blown, sealed and transferred to an aseptic chamber for opening, filling and sealing.
Bosch	Bottles/jars	Glass with PT metal cap or plastic with heat sealed membrane	Sterilisation of container and foil lidding material by hydrogen peroxide. PT cap by steam. Aseptic piston filler.

flowable vegetable product such as juice, purée or strained baby food could be handled by this method.

A number of systems are offered for blown plastic bottles, sterilisation of the container being effected during the hot extrusion of the plastic which reaches a temperature of 200 °C. Aseptic systems for glass bottles or jars are not readily available, although one is now being offered by Bosch.

Sachets and pouches. Systems for pouches either use chemical sterilisation of the pouch material or rely on the heat of extrusion of the film tubing for sterilisation (see Table 4.4). The bag-in-box applications are considered under the bulk packages section.

Cups. There are two types of system; those that form the containers from web directly on the machine and those that use pre-formed cups (see Table 4.5). Except for the Conoffast system, all rely on chemical sterilisation by hydrogen peroxide, accomplished in the former type by immersion of the web before the forming operation, and, in the latter, by spraying the inside of the container. The Conoffast system uses the heat of extrusion as a means of providing a

Table 4.4 Aseptic filling systems – sachets and pouches.

Manufacturer	Form	Material	Sterilising and filling method
Thimmonier	Pouch	Thermoplastic laminate	H_2O_2 bath plus UV irradiation used to sterilise the pouch material, which is filled and sealed in a sterile chamber
Prepac	Pouch	Thermoplastic laminate	H_2O_2 bath plus drying with filtered air
Berto	Pouch	Blown film polymer	Thermoplastic tube extrusion, assumed sterile inside. The outside is sterilised in H_2O_2. The sachet is cut, filled, and sealed in filtered air
Asepak	Pouch	Blown film polymer	Similar to above

sterile contact surface. This surface is exposed by stripping off the outer layer of the multilayered co-extrusion prior to forming the cups and ends.

Cartons. Two types of system are available; the web-fed machines which form the cartons directly on the machine and those that use pre-made cartons erected on the machine just prior to use (see Table 4.6). The former type, such as Tetra Brik, takes material directly from the reel and forms it continuously into a tube which is filled and then sealed below the product surface. This has the effect of eliminating the headspace, a procedure which is advantageous if the product is oxygen sensitive. On the other hand it excludes those products which contain particulates or fibrous material which would interfere with the integrity of the seal. Systems working with machine erected cartons, such as Combibloc, have found favour for this type of operation. However, Tetra can now provide machines which handle particulates in suspension. This is achieved satisfactorily by using two streams, one containing the solids in suspension and the other the sauce component. The latter flows continuously into the container while the former, pulsed by a pump synchronised with the web feed, is delivered at mid-point of the fill so that the seal area is avoided.

Bulk packages. Seasonal food is sometimes packed aseptically in bulk containers. The containers used may be metal drums, supported plastic bags or, for larger users, large aseptic tanks or rail-cars. The bag-in-box containers may be used for a variety of products suitable for catering or institutional use (see Table 4.7).

The Cherry-Burrell No-Bac Fifty Five and the Fran Rica Quadaseptic systems are widely used for the packing of tomato concentrate into drums.

In the Cherry-Burrell system, sterilisation, filling and sealing of the drum are carried out in an autoclave. The product is sterilised in a heat exchanger, partially cooled to 40 °C, and pumped to a surge tank where it is held under slight pressure from nitrogen gas. It is filled under vacuum into the drum, which has previously been sterilised in the autoclave by steam at 100 psig (690 kPa),

Table 4.5 Aseptic filling systems – thermoformed cups.

Manufacturer	Form	Material	Sterilising and filling method
Erca, Continental Can	Form-fill-seal cup	Thermoformable plastic co-extrusion; lid is from heat-sealable web	Delaminates multilayer coextrusions exposing sterile surfaces from which the cup and lid are formed; filled in sterile air tunnel
Benco	Form-fill-seal cup	Thermoformable plastic; lid is from heat-sealable web	The web for the body and the lid are immersed in H_2O_2; filled and sealed in a sterile tunnel
Plastimechanique-Formseal	Form-fill-seal cup	Thermoformable plastic; lid is from heat-sealable web	The web for the body and the lid are immersed in H_2O_2; filled and sealed in a sterile tunnel
Bosch-Servac	Form-fill-seal cup	Thermoformable plastic; lid is from heat-sealable web	The web for the body and the lid are immersed in H_2O_2; filled and sealed in a sterile tunnel
Micron-Formpak	Form-fill-seal cup	Thermoformable plastic; lid is from heat-sealable web	The web for the body and the lid are immersed in H_2O_2; filled and sealed in a sterile tunnel
Gasti	Preformed	Thermoplastic-based	H_2O_2 and hot-air treatment of cups and cut-out lids
Metal Box-Freshfill	Preformed	Thermoplastic-based	H_2O_2 and hot-air treatment of cups and lidding; sterile air filling chamber

Table 4.6 Aseptic filling systems – cartons.

Manufacturer	Form	Material	Sterilising and filling method
Tetra Pak Tetra Brik	Web-feb – precreased	Plastic/paper Alu foil laminate	H_2O_2 and heat; operates in an environment of H_2O_2 and steam
International paper	Web-fed – creased on machine	Plastic/paper Alu foil laminate	H_2O_2 and heat; operates in an environment of sterile air
PKL Combibloc	Machine-erected cartons	Plastic/paper Alu foil laminate	H_2O_2 and hot air; operates in sterile air
Liquipak	Machine-erected cartons	Plastic/paper Alu foil laminate	Synergistic H_2O_2 and UV light; operates in sterile air

Table 4.7 Aseptic filling systems–bulk packages.

Manufacturer	Form	Product type	Sterilising and filling method
Fran-Rica	Supported bags in sizes from 2.25 to 55 US galls. Closed by heat sealed membrane and screw overcap	High- and low-acid liquids. Particulates up to 0.5 inch	Gamma irradiation prior to shipment. The filling fitment sterilised by steam before fill cycle
Coloreed Liquid Packaging (Intasept)	Supported high-barrier bags. Closed by heat sealed membrane	High- and low-acid liquids including particulates	Gamma irradiation prior to shipment. The filling fitment sterilised by steam before fill cycle
Alfa-Laval/Akerlund & Rausing (StarAsept)	Bags made from various laminates and supported by drums, cardboard boxes and metal framed ply containers. Mechanical seal. Aseptic emptying possible. Wide size range	High- and low-acid product with particulates	Gamma irradiation prior to shipment. Steam sterilisation of filling spout prior to filling
Bowater Liquid Packaging Ltd (Bow-safe)	Bags made from various laminates and supported by drums or cardboard boxes. 3–25 litre size. Decap/recap seal	High- and low-acid product with particulates claimed	Gamma irradiation prior to shipment. Oxonia spray of spout prior to filling. Sterile air atmosphere
Corrugated Products Ltd (Liqui-Box)	Automatic web fed bags. Size 2–30 litres. Decap/recap seal.	Fruit and dairy-based liquids with and without particulates	Gamma irradiation prior to shipment. External sterilant spray in sterile air tunnel. Decap/recap in sterile zone
Scholle Corporation	Auto-fill 19A uses web fed bags with sizes of 3–24 litres. Decap/recap seal. Other machines are available		Gamma irradiation prior to shipment. External chlorine mist supplemented by steam to raise temperature. Sterile air zone
Tito Manzini	Bags 50–200 litres capacity with various supports. Heat sealed cap		Gamma irradiation prior to shipment. External treatment with iodophor. Filling through pierced membrane in cap

and a sterile closure is swaged in place. The system is suitable for packing a wide range of fruit and vegetable purées and concentrates. In the Fran Rica system, a sterilising nozzle introduces steam at 15 psig (104 kPa) into the drum. The system vents to the atmosphere for 1 min and 20 s to ensure removal of air. Following the venting cycle the container is pressurised to 18 psig for 40 s. The pressure is then reduced to 1.5 psig (10 kPa) and maintained at that pressure throughout the filling cycle. The filling tube is inserted to the bottom of the container and rises during the filling operation. After the product fill valve closes, the tube is retracted into the fill tube housing and rinsed with sterile condensate.

The Fran Rica system has been modified for the filling of acid products, such as fruit concentrates, into pre-sterilised flexible plastic bags supported in open top steel drums.

Different bag-in-box systems are in use. A common feature is pre-sterilisation of the sealed bags usually by treatment with gamma rays. These are opened under aseptic conditions just prior to filling. The early systems were designed for filling and containment of homogeneous acid products such as fruit purées. More recently systems have become available for low-acid products and some will handle particulates in suspension. A feature of these machines is the use of steam for sterilisation and protection of the filling gland or spout during the filling operation.

The Intasept system, used in the Coloreed machine, relies on a membrane of flexible material across the outer opening of the filling gland to preserve sterility after irradiation and prior to use. The area between this and the filling head is sterilised by steam before the membrane is punctured to admit the product. After filling and flushing of the seal area the bag is sealed by welding a pre-positioned membrane of barrier plastic across the base of the filling gland. A circular heat-seal head actually seals this membrane in place through the back of the bag. The back layers of the bag do not adhere to the membrane because they are of dissimilar material. Thus the dispensing fitments on the bag are isolated from the product, and it is claimed by this means, there is less oxygen permeation into the pack.

Another system using steam sterilisation of the filling area is StarAsept developed jointly by Alfa-Laval and Akerlund & Rausing. Based on a patented spout design, the StarAsept system is designed for large bags with a size to 1000 litres but it can be used for smaller bags of less than 25 litres. Aseptic emptying of the large bags is possible by provision of an optional emptying spout. Support for the bags depends upon their size and end use, but may be in the form of steel drums or cardboard boxes.

The smaller bags of up to 25 litres capacity are filled and closed on the HRI (StarAsept system) machine. They are sterilised before delivery and arrive in a paper-board box as a continuous tear-off web ready for feeding to the filler. The filler assembly comprises the operator's platform, control panel, the spout manipulation and filling device, and bag conveyor. After the first bag the filling procedure is fully automatic, although manual mode can be selected for

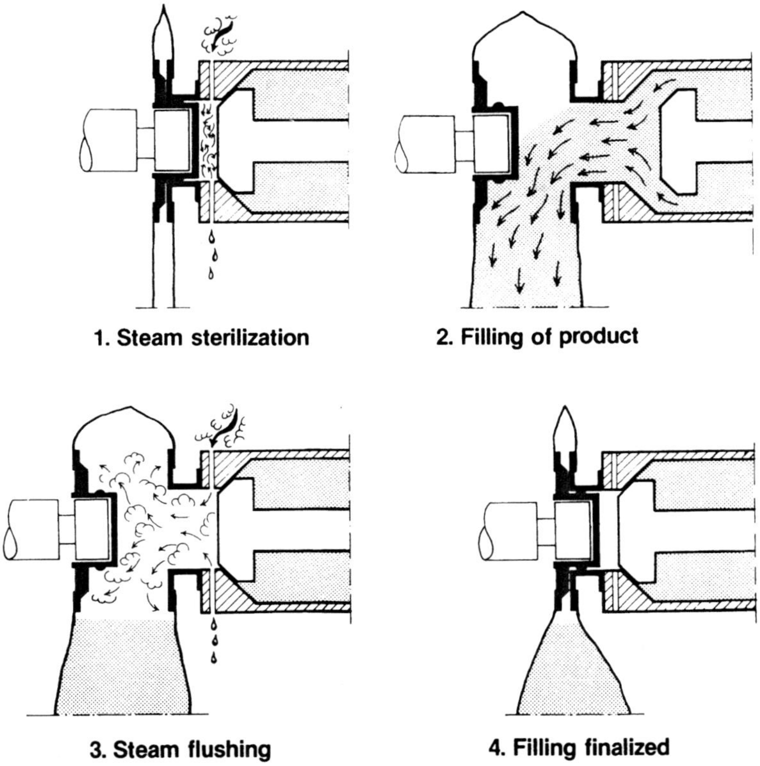

Figure 4.9 StarAsept filling sequence (courtesy Alfastar AB).

maintenance and test purposes. The filling sequence is shown in Figure 4.9.

The operator pulls the web from the box and places the first bag in position at the filling assembly. When the start button is pushed the spout is gripped, steam sterilised and opened for filling. After the first bag is filled and its spout closed it is released, and, as it slides down the conveyor, the next empty bag of the web is pulled into the filling position. As sterilisation of the second bag takes place, the first bag is cut free and drops through the conveyor into the waiting box underneath. The cycle is then repeated until the web is nearly depleted, at which time, a new web is spliced in place. Overall control is exercised by a microprocessor and sterilisation times and temperatures are monitored and recorded. The filling machine may be cleaned in conjunction with the aseptic tank by using a dummy spout which is placed in the filler valve.

4.7 Nutrient retention during thermal processing

A number of misleading comparisons have been made on the nutritive value of fresh and canned produce which have led to the common belief that fresh vegetables are invariably superior to processed vegetables. Lund (41) concluded from published data that nutrient losses appeared to be quite significant in the canning process. However, comparisons made against raw produce from retail outlets and prepared at home or in catering establishments in the usual manner do not always support this belief. Martin (42) compared processed vegetables with 'fresh' vegetables bought in retail stores and found that carotene, ascorbic acid, thiamin, riboflavin and niacin were considerably higher in the canned product. Differences in cultivars and loss of vitamins during transportation and storage could account for these findings. Since a major source of nutritive depletion is attributable to the draining of cooking liquors, it is likely that this will contribute to further loss from the fresh product as prepared for consumption. The National Food Processors Association has developed a nutritional labelling guideline based on surveys of American products, which gives nutritional information on canned products (Hagen *et al.* (43)).

In spite of the above reservations it cannot be disputed that the canning process does reduce the nutritive value. Vitamin losses occur during commercial processing, very often in the blanching as well as during sterilisation. Vitamin A is relatively stable, due to its temperature stability and insolubility in water, and survives the canning process quite well. Vitamin B_1 (thiamin) is heat labile and losses occur in blanching and in sterilisation. These can be reduced by using an HTST process in conjunction with aseptic filling. Everson *et al.* (44) reported that the thiamin retention was increased from 60% to more than 80% by this means. Vitamin B_2 (riboflavin) is heat-stable but, due to its solubility in water, losses occur during blanching and in the drainage liquors from the can. Niacin behaves in a similar manner. Vitamin C (ascorbic acid) is relatively unstable to heat, oxygen and light and is water soluble. Blanching reduces the level of vitamin C by 25–35% and there is a further loss during and immediately after sterilisation, as oxygen from dissolved gases and from the headspace is absorbed.

4.8 Storage of canned products

Although canned foods are one of the most stable food products organoleptic and nutrient changes occur during storage. The extent and rate of these changes depend upon the temperature of storage, the nature of the packaging, the product characteristics and the extent and severity of the heat treatment which was applied to effect sterilisation. This is well illustrated by green vegetables or vegetable purées sterilised by an HTST process and packed aseptically. When first packed, such products are much higher in sensory quality than similar products produced by regular canning methods, particularly in respect of the retention of green colour. However, during storage at temperatures of 21 °C or higher, this quality differential is eroded so that the difference is negligible after about six months. Storage at low temperatures is necessary to delay this deterioration.

Table 4.8 Maximum storage temperatures (°F) for canned foods to ensure not more than 10% loss of a selected vitamin for 12, 18 and 24 months' storage (from Kramer (45)).

	Ascorbic aicd (vitamin C)			Thiamin (vitamin B1)			Carotene (vitamin A precursor)			Niacin			Riboflavin (vitamin B2)		
Canned product	12	18	24	12	18	24	12	18	24	12	18	24	12	18	24
Apricots	76	68	60	32	–	–	70	62	54	–	–	–	–	–	–
Asparagus	74	65	58	45	38	32	72	58	45	b	b	b	57	50	45
Beans, green	60	45	32	57	45	32	–	–	–	b	b	b	57	50	45
Beans, lima	74	60	45	32	–	–	–	–	–	c	c	c	65	40	32
Carrots	–	–	–	–	–	–	>80	>80	>80	–	–	–	–	–	–
Corn, sweet	71	60	50	64	58	53	58	66	75	d	d	d	70	55	48
Frankfurters and beans	–	–	–	56	47	41	–	–	–	–	–	–	–	–	–
Grapefruit segments	60	45	32	–	–	–	–	–	–	–	–	–	–	–	–
Peaches	65	50	40	76	80	84	70	40	32	c	c	c	–	–	–
Peas	74	66	52	62	56	50	72	58	45	b	b	b	55	50	45
Plums	–	–	–	–	–	–	>80	>80	>80	d	d	d	f	f	f
Pineapple slices	68	54	32	80	80	80	–	–	–	–	–	–	–	–	–
Spinach	68	60	52	63	57	49	46	41	38	c	c	c	63	46	32
Tomatoes	68	60	45	70	60	45	a	a	a	e	e	e	d	d	d

a 10% loss in 17 months, no temperature effect
b 20% loss in 12 months, 15% in 24 months, no temperature effect
c No loss
d 10% loss, no time or temperature effect
e 8% loss in 12 months, 13% in 24 months, no temperature effect
f 20% loss, no time or temperature effect

Small losses of vitamins occur during storage although the extent is dependent upon temperature. Kramer (45), using data from the literature, predicted the storage temperatures that could be tolerated by canned foods if no more than 10% of a specified nutrient was to be lost during a stated storage period (Table 4.8). Lee *et al.* (46) detected minor losses of thiamin and ascorbic acid in canned beans and broccoli after storage for six months at 0 and 5 °C and considerably greater losses at 21 and 35 °C. Packaging has an effect on the rate and extent of ascorbic acid loss particularly for acid products. Unlacquered cans remove oxygen from the headspace faster than lacquered cans, glass or thermoplastic containers resulting in better ascorbic acid retention in products packed in such containers. Continuing ascorbic acid loss may be expected during storage in containers which permit a slow ingress of oxygen.

Textural quality is important in canned vegetables with firmness and crispness rated as positive attributes. Loss of these characteristics during processing is one of the main criticisms of canned vegetables and methods which permit the application of HTST processes, such as pouch packaging or aseptic packaging, generally lead to better textures. Further textural changes may occur during storage. Shannon *et al.* (47) evaluated the firmness of canned carrots and red beets after they had been stored for 600 days at 18 and 38 °C. Carrots firmed slightly but the storage temperature had no effect. However, there was a significant decrease in firmness of the red beets with time and the rate of decrease was four times as rapid at 38 °C. Changes, such as clumping of peas and beans, may occur due to retrogradation of the starch and these are generally more apparent in products stored at low temperatures.

The flavour of canned vegetables may change during storage, particularly at high temperatures. Hayakawa *et al.* (48) studied the sensory quality changes of asparagus, corn kernels, peas and apple slices after storage at different temperatures ranging from −29 to 38 °C for periods up to one year. In general the products stored at 38 °C deteriorated much faster than those stored at the lower temperatures. Storage at 2, 4.5 and 21 °C did not result in severe loss of quality over the entire storage period. For products high in sugar, such as fruits canned in syrup, browning reactions have been known to give rise to burned or bitter flavours during storage, particularly at high temperatures. However, in general, canned vegetables store well if the temperature is maintained at a reasonable level, say below 21 °C, and provide a valuable addition to the diet.

References

1. Esty, J.R. and Meyer, K.F., *J. Infect. Dis.* **31** (1922), 650.
2. Bigelow, W.D. and Cameron, E.J., *Ind. Eng. Chem.* **24** (1932), 655.
3. Cameron, E.J. and Esty, J.R., *Food Res.* **5** (1940), 549.
4. Holdsworth, S.D., *The Preservation of Fruit and Vegetable Food Products, Sci. in Hort. Series*, Macmillan, London, 1983.

5. Ball, C.O., *Food Res.* **3** (1938), 13.
6. Bendix, G.H., Heberlein, D.G., Ptak, L.R. and Clifcorn, L.E., *Food Res.* **16** (1951), 494.
7. Feliciotti, E. and Esselen, W.B., *Food Technol.* **11** (1957), 77.
8. Gupte, S.M., El-Bisi, H.M. and Francis, F.J., *J. Food Sci.* **29** (1964), 379.
9. Hayakawa, K. and Timbers, G.E., *J. Food Sci.* **42** (1977), 778.
10. Hayakawa, K., *Can. Inst. of Food Technol. J.* **2** (1969), 165.
11. Mansfield, T., in *Proc. of 1st Int. Congress of Food Sci. and Technol.* London, 18–21 Sept, 1962, Vol. IV, *Manufacture and Distribution of Foods* (ed. Leitch J.M.), Gordon and Breach, London, 1965, p. 311.
12. Guyer, R.B. and Holmquist, J.W., *Food Technol. Champaign* **8** (1954), 547.
13. Adams, J.B., *J. Food Technol.* **13** (1978), 281.
14. Ings, J., *Metals Australia* (Aug. 1976), 150.
15. Ray, A., *Food Process.* **36** (1) (1975), 64.
16. Robe, K., *Food Process.* **34** (1) (1973), 12.
17. Ralls, J.W., *Proc. Res. Highlights. Natl. Canners' Assoc.*, Western Res. Lab., 1971, 105.
18. Lazar, M.E., Lund, D.B. and Dietrich, W.C., *Food Technol.* **25** (1971), 684.
19. National Food Processors Association, *Thermal Processes for Low-acid Foods in Metal Containers* (12th edn), Bulletin 26-L, NFPA, Washington DC 20005, 1982.
20. National Food Processors Association, *Thermal Processes for Low-acid Foods in Glass Containers* (5th edn), Bulletin 30-L, NFPA, Washington DC 20005, 1984.
21. Beauvais, M., *Rev. Conserve* **19** (7) (1964), 209.
22. Casimir, D.J., *CSIRO, Food Res. Quart.* **35** (1975), 34.
23. Leonard, S., Merson, R.L., Marsh, G.L., York, G.K., Heil, J.R. and Wolcott, T., *J. Food Sci.* **40** (1975), 246.
24. Leonard, S., Marsh, G.L., Merson, R.L., York, G.K., Heil, J.R., Fryer, S., Wolcott, T. and Ansar, A., *J. Food Sci.* **40** (1975) 250.
25. Leonard, S., Marsh, G.L., Merson, R.L., York, G.K., Buhlert, J.E., Heil, J.R. and Wolcott, T., *J. Food. Sci.* **40** (1975), 254.
26. Richardson, P.S., *Int. J. Food Sci. Technol.* **22** (1987), 3.
27. Lopez, A., *A Complete Course in Canning and Related Processes*, Book 1–*Basic Information on Canning* (12th edn), The Canning Trade Inc., Baltimore, Maryland, USA, 1987.
28. Benjamin, H.A. and Ecklund, O.F., *Food Ind.* **18** (4) (1946), 523.
29. Kenyon, E.M., Westcott, D.E., La Casse, P. and Gould, J.W., *J. Food Sci.* **36** (1971), 289.
30. Ayoub, J.A., Berkowitz, D., Kenyon, E.M. and Wadsworth, C.K., *J. Food Sci.* **39** (1974), 309.
31. Ohlsson, T., *Int. Seminar on New Trends in Aseptic Processing and Packaging of Foodstuffs*, Munich, 22–23 October, 1987.
32. Ball, C.O. and Olson, F.C.W., *Sterilization in Food Technology*, McGraw-Hill, New York, 1957.
33. Ziemba, J.V., *Food Eng.* **36** (3) (1964), 122.
34. Hersom, A.C., *Food Rev. Int.* **1** (1985) 215.
35. Bates, R.P. and Koburger, J.A., *Procs. of the Florida State Horticultural Soc.* **87** (1974), 245.
36. Stephens, T.S., Saldana, G., Brown, H.E. and Griffiths, F.P., *J. Food Sci.* **36** (1971), 36.
37. Hersom, A.C. and Shore, D.T., *Food Technol.* **35** (5) (1981), 53.
38. Grikitis, K., *Food Process.* **58** (6) (1989), 12.
39. Havighorst, C.R., *Food Ind.* **23** (1951), 72.
40. Anon, *Food Engng.* **23** (1951), 68.
41. Lund, D.B., Effects of heat processing on nutrients in *Nutritional Evaluation of Food Processing* (eds Karmas, E. and Harris, R.S.), Van Nostrand Reinhold, New York, 1987, p. 319.
42. Martin, S., *Quick Frozen Foods* (November 1977), 34.
43. Hagen, R.E., Elkins, E.R. and Farrow, R.P., *Hort. Sci.* **14** (1979), 251.
44. Everson, G.J., Chang, J., Leonard, S., Luh, B.S. and Simone, M., *Food Technol.* **18** (1964), 84.
45. Kramer, A., *Food Technol.* **28** (1) (1974), 50.
46. Lee, W.N., Kramer, A. and Onishi, Y., *J. Food Qual.* **2** (1979), 257.
47. Shannon, S., Bourne, M.C. and Rao, M.A., *J. Food Sci.* **42** (1977), 1674.
48. Hayakawa, K., Timbers, G.E. and Stier, E.F., *J. Food Sci.* **43** (1978), 944.

5 Freezing

DAVID REID

5.1 Introduction

Many vegetable are seasonal crops, abundant at certain times of the year, and not readily available at other times of the year. In order to extend availability all year round, methods of preservation are needed. These methods may alter the characteristics of the vegetable; some to a small extent, some to a large extent. They may also be effective over different time periods. Some extend shelf-life by only a few weeks, some give a shelf-life of many months. The changes in the tissue characteristics consequent upon the preservation technique may be more marked where long storage lives are required. For long-term storage, extending from months to years, canning (in which product is placed in a hermetically sealed container, and then heated sufficiently to achieve sterility) and freezing (in which product is cooled down to temperatures below $-20\,^{\circ}C$ and maintained at these temperatures) are the primary preservation techniques. Freezing is often preferred over canning, because the alterations to the tissues, such as softening and colour changes, may be less marked. Preservation by freezing makes use of the phenomenon that rates of chemical change are generally reduced at lower temperatures.

Rates of physical change, too, may be reduced at sufficiently low temperature, so that a product may be stored at low temperatures for extended periods without there being too much change in its properties compared to those it had at the commencement of the storage period. The freezing process itself, however, may result in change. There are commodities, such as lettuce, which are so altered by freezing as to make the method unsuitable for preservation of that particular commodity. It is generally true that the lower the temperature of storage, the slower the rate of change in properties.

Because freezing involves the separation of some of the water in the product as ice, with consequent increase in the concentrations in the unfrozen phase, and potential crystallisation or precipitation of other components, the rate of deterioration may be affected by factors other than the temperature linked slowdown. Higher solute concentrations may induce increased rates of change in some chemical processes (1). Also, low temperatures may enhance the rates of some crystallisation processes. The rates of change in frozen systems could then be a complex function of temperature. There may be a temperature region with enhanced rates of change just below $0\,^{\circ}C$ in some systems. However, at

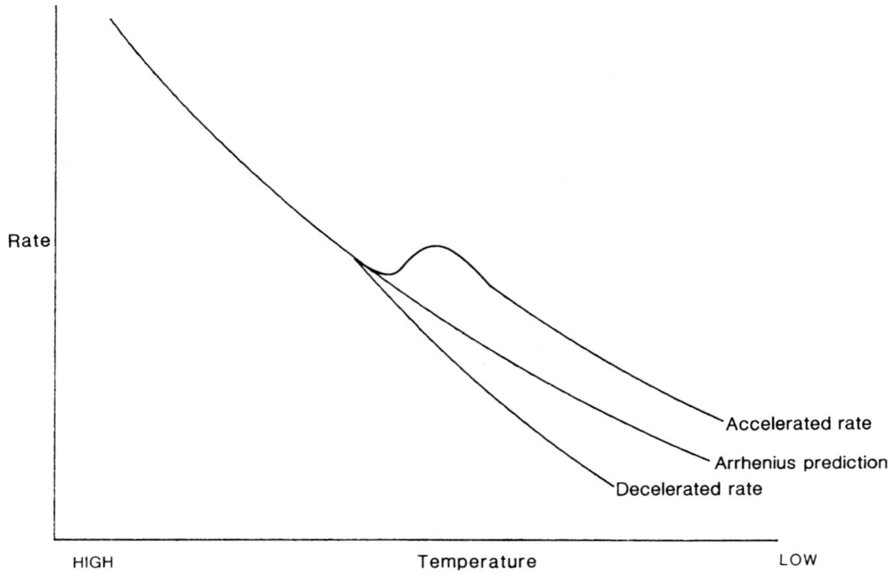

Figure 5.1 Rates of reaction as they may be affected by freezing.

sufficiently low temperatures, rates of change will reduce. This is illustrated schematically in Figure 5.1. The upper line illustrates rate enhancement due to freezing. The lower line illustrates rate inhibition due to freezing. The centre line represents the rate to be expected on the predictions of the Arrhenius rate expression. The Arrhenius expression is inappropriate for use in frozen systems, as it does not allow for the complex changes which accompany the separation of crystalline ice, but assumes the system stays one uniform phase. It has recently been shown that, in systems containing ice, the temperature dependence of the kinetics in the unfrozen phase may better be described by an equation, known as the Williams–Landel–Ferry equation, which describes the temperature dependence of kinetics in rubbery systems (18).

5.2. The freezing process

To maintain vegetables in frozen storage, first they must be frozen. Freezing requires the removal of heat, and so it will be necessary to discuss the means by which heat can be removed from these materials in order to freeze them. Different methods for heat removal will result in different rates of heat removal, and this in turn will result in different patterns of freezing. The characteristics of vegetables, too (such as size, internal structures, cell wall thickness), will affect the patterns of freezing. In particular, the cellular nature of the tissue, and the characteristics of the cell wall, can affect the movements of water within the tissue during the freezing process (2, 3), which in turn can also be an important factor affecting the location and size of ice crystals in the frozen tissue.

The rate of heat removal from the tissue will depend upon several factors (4). These are described in Sections 5.2.1 and 5.2.2.

5.2.1 *Heat exchange materials*

A material with high thermal conductivity, or with high volumetric heat capacity, will be more effective at removing heat than one of low thermal conductivity or low volumetric heat capacity under the same thermal contact conditions. The temperature of the heat exchange medium will also be important; the lower the temperature, the greater the driving force for heat exchange, and the more rapid the heat exchange. Note that a material with low heat capacity will increase in temperature quite rapidly with a relatively small increase in heat content, and so it might be difficult to maintain low temperature in a heat exchange medium with this characteristic. Air has a low volumetric heat capacity, and a low thermal conductivity, and so for air to be an effective heat exchange material, it must be at a low temperature, and the surface of the product must be brought into contact with a large volume of air in a short time, in order to remove enough heat and at the same time retain the low air temperature. Hence still air is an ineffective heat exchange medium, while high velocity, very cold air can be very effective. Cold metal can be a very effective heat exchange material, because of the ability to remove heat rapidly through thermal conductivity, but this requires that there be effective contact between the surface of the product and the metal surface. Anything in the contact between product and metal which acts as an insulator will reduce the efficiency of heat exchange. Another requirement is that the metal must be able to transfer the heat it collects into an efficient heat sink.

Liquids can be a good heat exchange medium, because of their volumetric heat capacity. Low-temperature liquid can be used effectively for freezing. Cold brine is an example. Immersion gives good thermal contact with the product surface. To improve on volumetric heat capacity, a heat exchange medium which undergoes phase change can be useful, because the latent heat required for the phase change can be drawn from the product, without there being an increase in the temperature of the heat exchange medium. The low-temperature driving force for heat exchange can therefore be maintained. Materials which undergo phase change at low temperatures and have commonly been employed are liquid nitrogen, solid carbon dioxide, and liquid freon. Liquid nitrogen, which boils at $-194\,^{\circ}C$ can be a very effective coolant. In addition to the enthalpy of the phase change, there is the possibility of gaseous heat exchange with very cold gas which can be moved at high velocity. Solid carbon dioxide, which removes heat as it sublimes at about $-80\,^{\circ}C$, also produces a cold gas. Liquid chlorofluorocarbons, for example, Freon12, which boils around $-40\,^{\circ}C$, can also be very effective heat exchangers, due to excellent thermal contact. However, the use of chlorofluorocarbons as liquid freezants are no longer recommended, due to the environmental consequences of their escape to the atmosphere. The methods of heat exchange using these different media will be considered further in relation to the practical aspects of freezing.

5.2.2 *Characteristics of the vegetable*

The size of the individual item to be frozen, that is, the area of the heat exchange surface as it relates to the volume of the object from which heat is to be removed, will be important to the time required for freezing. It will also affect the rate at which the temperature changes within the object, and the rate at which heat is removed at any point within the object. This is important because the rate of heat removal at a point is an important factor controlling the freezing behaviour at that point (5).

Vegetable materials to be frozen range in size from peas and diced carrots on the one hand to whole ears of corn on the other. The heat exchange characteristics at these two extremes are very different, so that the methods of handling peas and diced carrots are quite different from the methods of handling cob corn. This will be amplified in relation to the freezing methodology for individual vegetable products.

The freezing process is complex. The removal of heat from an aqueous system causes the temperature to reduce. This by itself is insufficient for freezing. A phenomenon known as undercooling can occur. Before ice can separate out, a seeding process, known as nucleation, must initiate crystallisation. Ice crystals then propagate from these seeds. In Figure 5.2 the pattern of freezing is illustrated by showing the time–temperature history at a location within the object which can experience undercooling. As heat is removed, the temperature drops below the freezing point at A, undercooling persists to B, where nucleation initiates growth. The release of latent heat then brings the temperature up to C. Further cooling follows the profile shown, and the amount of ice increases. These processes of nucleation and propagation are very much affected by the rate of heat removal. Rapid removal of heat allows a large undercooling to develop, and nucleation takes place at many centres. Small ice crystals form. A slow removal of heat allows for much less undercooling. Fewer nuclei lead to fewer, larger ice crystals. One of the factors which is important to this effect of rate of heat removal is the rate of propagation of ice. The rate of propagation is dependent upon the undercooling, and increases with increasing undercooling. Hence a condition which requires rapid propagation naturally leads to greater undercooling, and therefore more abundant nucleation, and more, smaller crystals (5, 6).

In an intact tissue system, other factors have to be considered. The cell wall and cell membrane can be an effective barrier to the propagation of ice. It is therefore possible to nucleate ice in the intercellular space, and have this ice

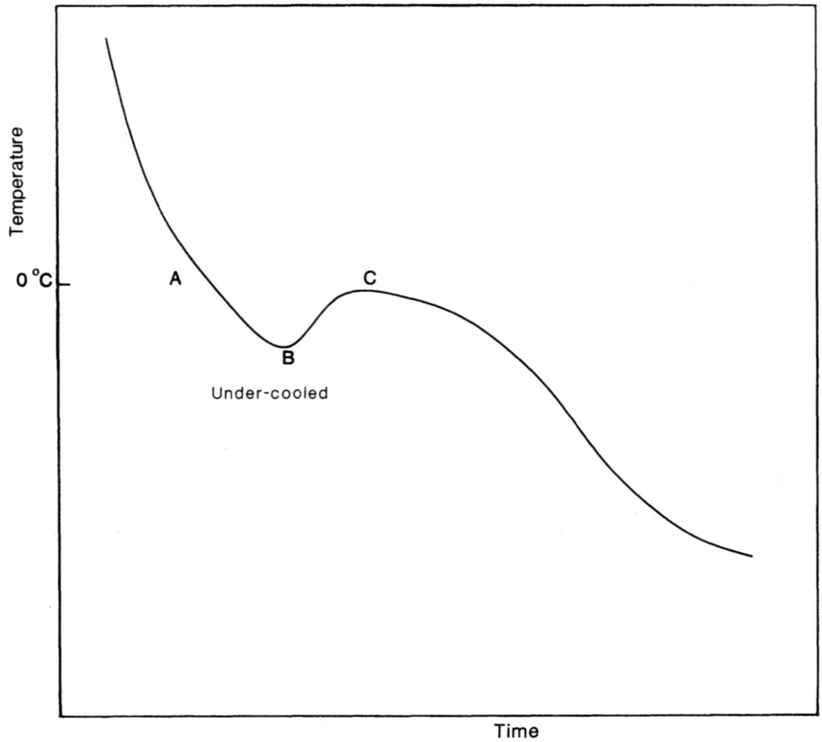

Figure 5.2 A typical freezing curve showing undercooling.

propagate in this space, but to have the cell contents unfrozen. Should this happen, there is an osmotic driving force for dehydration of the cell, water transferring to the extracellular space containing the ice, and a more highly concentrated unfrozen liquid phase. The permeability of the cell wall will influence the rate at which this dehydration can take place. If the extracellular medium continues to cool, and therefore concentrate its fluid component, continued water removal will be required to maintain osmotic balance. Should this not occur, or not occur sufficiently rapidly, the cell will undercool. The extent of undercooling will depend both on the rate of change of temperature of the extracellular medium surrounding the cell, and the rate of osmotic dehydration permitted by the cell wall. At some critical undercooling, the cell may nucleate ice internally, and freeze. Hence depending upon both the cooling conditions, and the specific characteristics of the cell, the cells may dehydrate, or they may freeze internally. Internal freezing under these conditions often causes major damage to the functionalities of the cell. As already indicated, the rate of heat removal will be affected both by the freezing method, and by the size of the tissue being frozen. Rapid freezing of peas is easily achieved. Rapid freezing of cob corn is not practical.

The texture and initial colour of rapidly frozen raw tissue are often superior to those of the slowly frozen raw tissue. However, in many cases it is found that there is generation of off-flavours, off-odours and off-colours during extended frozen storage (7, 8). These changes may be more rapid in the rapidly frozen systems, but they are seen too in the slowly frozen systems. In other cases the product appears to have adequate stability. The texture of slowly frozen tissues is usually inferior, due to the relocation of the water within the tissue.

5.3 Blanching

Where marked flavour changes occur, pre-treatment of the tissue to yield a more stable product is required. The changes are a consequence of the disruption of the internal cell structures, leading to uncontrolled enzyme catalysed reaction. Some of the products of these reactions produce the off-flavours and off-odours. Enzyme inactivation can be achieved by blanching, a controlled heat treatment. Three principal types of blancher exist, where heat transfer is through hot water, through steam, and through hot gases respectively (9, 10). Blancher designs have evolved to give more precise control of the thermal treatment, and to cut energy consumption. Often the requirements of precise control and of energy efficiency are complementary. It is important that the flow of product through a blancher be regulated in such a way that the thermal treatment of individual vegetables be consistent. Care has to be taken that there is uniform heat transfer. A thick layer of product might result in very different thermal histories for material near the surface of the layer, and material deep within the layer. This is important, since the extent of enzyme inactivation will depend on the extent of heat treatment. The size of product to be blanched must also be taken into account, because this will affect the rate of heat transfer, and the time to achieve a target temperature at the centre of the product.

Because blanching produces secondary changes, such as tissue softening and pigment degradation, in addition to enzyme inactivation, it is necessary to control the blanch, and prevent there being too extensive a heat treatment, as this will result in an inferior final product. It is also necessary to ensure that the appropriate enzyme is inactivated, as residual enzyme activity will cause deterioration during storage. Traditionally, tests have been evolved to indicate residual enzyme activity, and blanch conditions necessary for inactivation have been determined and tabulated (11, 12). The enzyme most often used as a marker has been peroxidase, which is of relatively high thermal stability, is easy to assay, and is present in all vegetables. Because peroxidase has high stability, a blanch sufficient to inactivate peroxidase is considered to be adequate to inactivate all significant enzymes. Sometimes catalase has been used as an indicator enzyme. This enzyme is easy to test for, too, and has lower stability, so that a less severe thermal treatment is adequate for inactivation.

One criticism of the use of catalase as marker is that it is not present in significant quantities in many vegetables.

The use of peroxidase as marker enzyme, however, may result in the use of more severe blanching conditions than might really be necessary (13, 14). For example, in peas and green beans the enzyme responsible for most of the off-flavour and off-odour generation is lipoxygenase (14). This enzyme is less stable than peroxidase, so a shorter or a cooler blanch might suffice for inactivation. It would, of course, be necessary to have a reliable test for residual lipoxygenase activity for use in quality control. Such a test has been developed, and indeed a shorter blanch than is required to inactivate peroxidase has been shown to produce a stable product. New recommendations for blanching conditions might be forthcoming. With the more precise control of blanching conditions which is possible in modern equipment, less severe thermal processing may safely be applied, so that better consistency in the product fed to the freezer might be achieved.

5.4 Industrial freezing

The industrial process of vegetable freezing consists of several steps. First the vegetable is harvested and transported to the processing plant. This should be done as expeditiously as possible, because quality will be lost if product is held under inappropriate conditions. The material will have to be cleaned, sorted and graded. This may be done at any stage between farm and entry to the process line. For different commodities different methods will be used. Grading may be on the basis of size, colour, maturity, or many other factors. After appropriate selection and grading of the raw material the main processing can be commenced. Further cleaning may be necessary. The material may have to be peeled. Some large vegetables may need to be cut up. For some vegetables, blanching will be required, so they will be transported through the blancher, and cooled prior to further processing. The blanching conditions will vary according to the size and characteristics of the vegetable.

As already stated, freezing requires the removal of heat from the product to bring the temperature down to the temperature of storage, or below. This heat transfer can be achieved through the use of the different types of heat exchange media identified earlier. For example, heat can be removed by contacting the product with cold air. Some means is required, of course, to generate the supply of cold air. This is achieved through mechanical refrigeration, which is used to cool heat exchange coils through which the air can flow, and be cooled down to the temperature required for its utilisation in the freezer.

Mechanical refrigeration systems are made up of four elements: a compressor, a condenser, an expansion valve and an evaporator (Figure 5.3). Heat is removed from the region of low temperature by the evaporation of the working fluid, which is under low pressure, and is part liquid. The vapour flows to the compressor, which increases the pressure with resultant increase in the

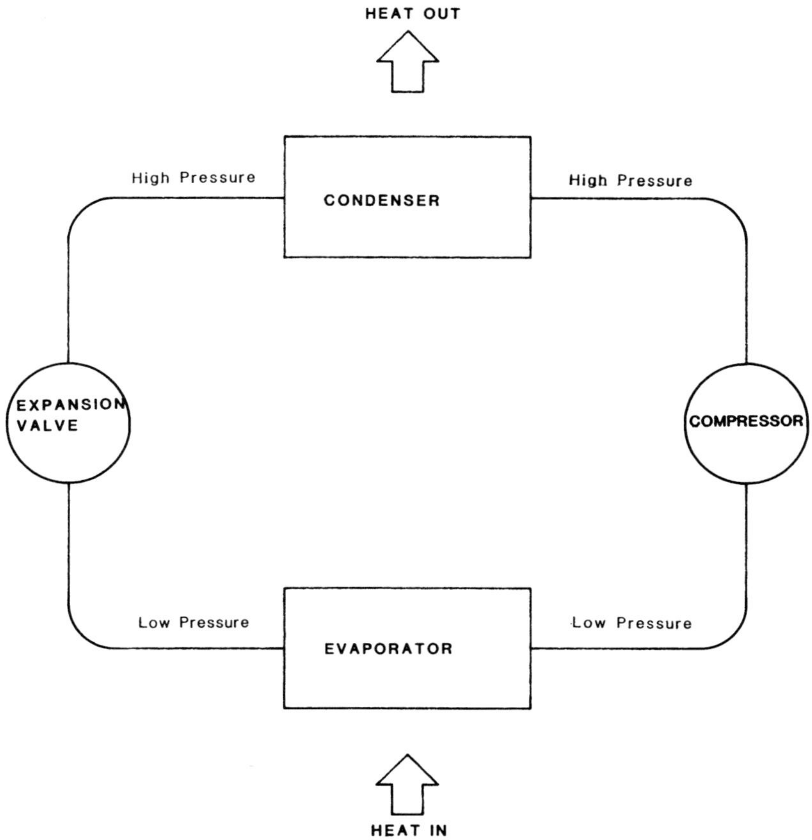

Figure 5.3 The mechanical refrigeration cycle (schematic).

temperature of the working fluid. Heat is now removed from the working fluid and rejected to the external environment in the condenser, at which stage the fluid becomes liquid. This liquid may now pass through the expansion valve to the low-pressure side. The temperature decreases, and the cycle starts again with the reduced pressure liquid boiling in the evaporator, and removing heat as required.

Several refrigerants are in common use (14, 15). These include some chlorofluorocarbons, and ammonia. Important properties required of effective refrigerants include:

(1) A low boiling point – to have a low evaporator temperature without requiring too low an evaporator pressure.
(2) A high latent heat of evaporation – to minimise the amount of refrigerant which has to evaporate to give the required cooling effect.
(3) A dense vapour – this will keep down the size of compressor.

(4) A practical pressure at the temperature required for heat rejection in the condenser. The environmental temperature of the condenser has to be considered, for example, an air-cooled condenser in California could require a condenser temperature of above 40 °C; a water-cooled condenser might operate at 25 °C or below. Too high a condenser pressure can cause engineering problems.
(5) A low toxicity – in case of leaks. In the case of ammonia, ease of leak detection enables us to disregard this requirement.
(6) A low flammability.
(7) A low cost.

No one refrigerant can meet all requirements, so the refrigerants used by industry are chosen on the basis of practicality.

As indicated, mechanical refrigeration can be used to cool coils across which air is passed. These evaporator coils can be operated at a range of temperatures to give the air temperature desired. The air velocity across the coils also can be varied as required. The chilled air may then be channelled to the freezing equipment. Depending upon factors such as air temperature, air velocity and product size, the freezing time for product within the freezer will vary. Some freezers which use air as a heat exchange medium consist of chambers into which product is loaded and held in contact with cold air. This may be a batch procedure, or carriers loaded with product may be moved through in a semi-continuous pattern. In other freezers, conveyers or other means may be used to transport product through a zone of cold air. The air temperature, product path, and residence time can all be varied to suit the process. Chilled air is not the only effective heat transfer medium, a variety of alternatives exist. These may be more appropriate in some circumstances. In one alternative, the product is contacted by cooled metal plates. These plates are cooled internally, either by circulation of a secondary heat exchange fluid, which also circulates through the heat exchange coils of the mechanical refrigeration unit, or by direct expansion of the refrigerant within the plates. The effectiveness of plate freezers is critically dependent upon the efficiency of heat exchange between product and cooled plate. Where there is good contact between the product and the plates this conduction method can be a very effective method of freezing.

Immersion freezers use a liquid as heat transfer medium. The most commonly used immersion liquid is a cooled brine. This brine is cooled by circulation through the evaporator coils of the mechanical refrigeration system. Heat transfer is good. As stated earlier, liquid chlorofluorocarbon has been used as an immersion liquid, but environmental concerns will probably result in discontinuation of this use. Cryogenic fluids may also be used in immersion heat exchange to benefit from the latent heat factor, and at the same time make it unnecessary for the plant to have mechanical refrigeration.

Three primary means are available for delivering product to the point at which it will be in contact with the selected heat transfer agent. These are: (1) batch-wise, with the processing of each batch completed before the introduction of the next batch; (2) semi-continuous, with the product delivered and removed in batches, but with each batch requiring several time cycles before removal; and (3) continuous, with a continuous product feed and removal. The characteristics of each method will be considered separately.

5.4.1 *Batch freezers*

Figure 5.4 schematically shows methods of delivering material into the freezing environment in a batch system. The freezing environments utilised include blast-air freezers and plate freezers.

5.4.1.1 *Blast-air freezer.* The blast air batch freezer is probably the simplest type of freezer available for use with vegetables (Figure 5.5). It consists of a

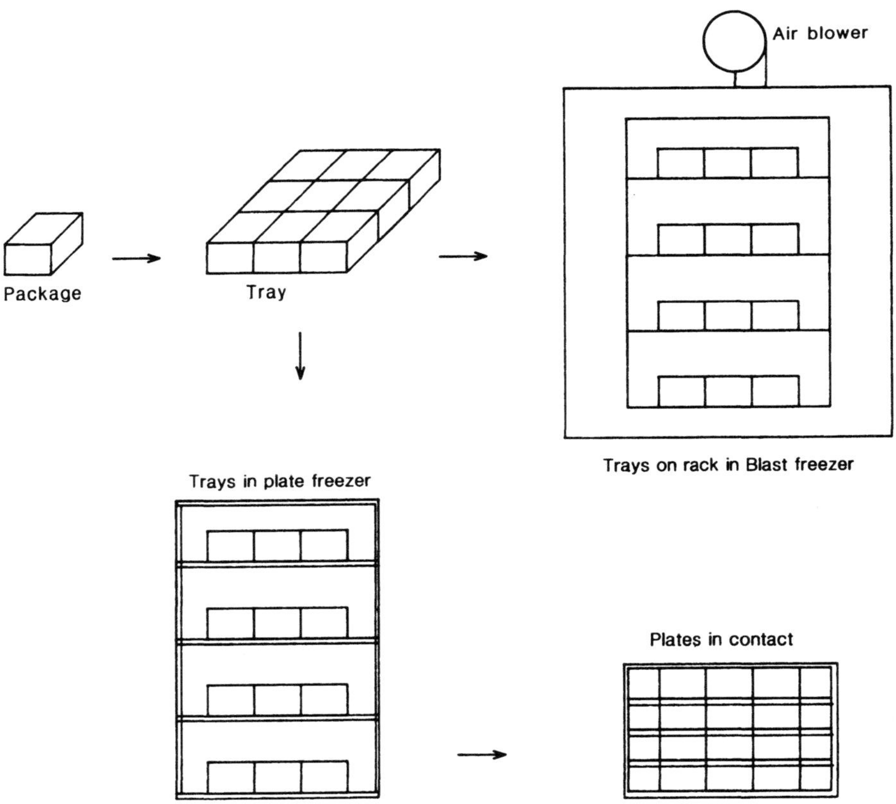

Figure 5.4 Schematic illustration of batch methods of freezing.

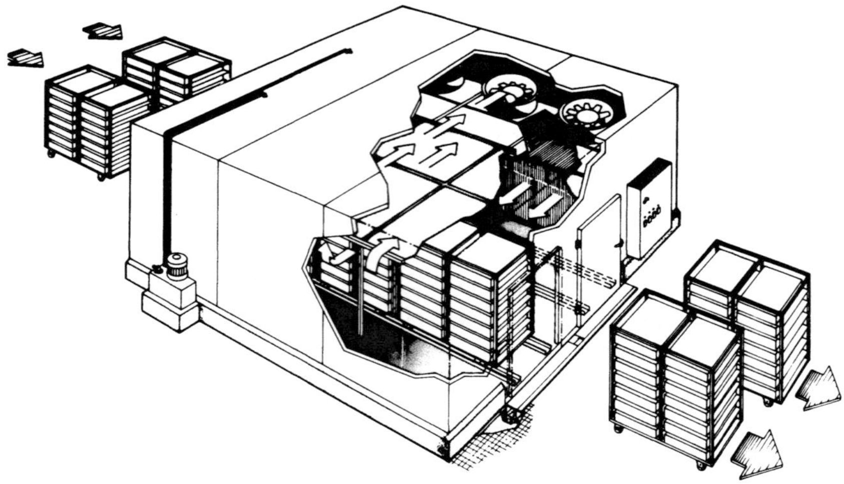

Figure 5.5 A blast freezer with trolleys (courtesy Frigoscandia Contracting AB).

freezer chamber, and some means of transferring product in to and out from the chamber, often on wheeled racks. Racks may have to be loaded and unloaded by hand. Produce can be loaded on to trays, and these trays placed in racks. The racks are then loaded into the chamber. The method is particularly suited to small-scale production, but tends to result in slower freezing. Product may stand around for some time before freezing commences. Depending upon the design of the chamber, and of the racks, air distribution may be fairly uniform, or it may be uneven, and therefore poor. Great care has to be taken to prevent air from channelling past the product without there being sufficient contact for effective cooling. A properly designed batch freezing chamber, with well-sited fans, can have adequate performance. However, in some cases, trays of product loaded upon racks are simply placed in large refrigerated rooms which have not been designed for the purpose. This is one of the poorest procedures, as airflow is not well controlled. It is even worse if the room is in use for storage, because the heat load from the unfrozen product could cause unacceptable warming. Many older freezing installations still have blast-air batch freezers. The air is cooled by circulating it across the refrigeration coils which are cooled by conventional mechanical refrigeration. Properly designed and maintained they can produce adequate results for products where freezing rates are not critical.

5.4.1.2 *Batch-plate freezers.* Many plate freezers are operated as batch freezers. Product is loaded onto trays, and placed in the freezer. The plates are brought into contact with the loaded trays, and freezing commences. Most often, the product is pre-packaged in boxes before loading, as regular-shaped packages are best suited to making contact with the plates. Thermal contact

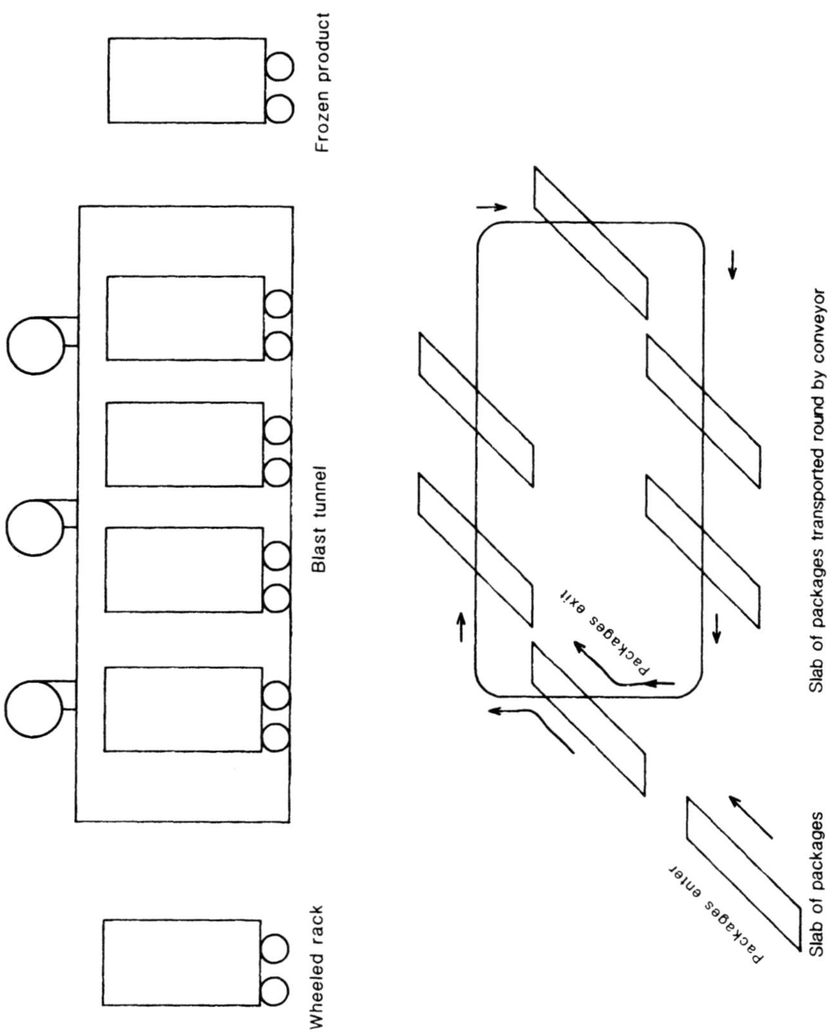

Figure 5.6 Schematic illustration of semi-continuous methods of freezing.

within the boxes may not be good, so this is not a recommended method for freezing most vegetables, though it works well for purées, and limp leafy vegetables such as spinach, where it is possible to produce a block of product which fills the rectangular package without much void space.

5.4.2 *Semi-continuous freezers*

In semi-continuous freezers, batches of product are loaded in at one end of the freezer, they travel through the freezer, and batches are removed at the exit. Racks of product can move through on a rail system, or some means can be used to collect a product batch which is introduced all at once. Some of the procedures are illustrated schematically in Figure 5.6. Semi-continuous systems can be designed for air-blast freezers and for plate freezers (Figure 5.7), and share many of their characteristics. Throughput is increased compared to the simple batch freezer. For longer freezing times, residence time can be increased by increasing the length of the chamber, hence increasing the number of racks. This will not affect throughput.

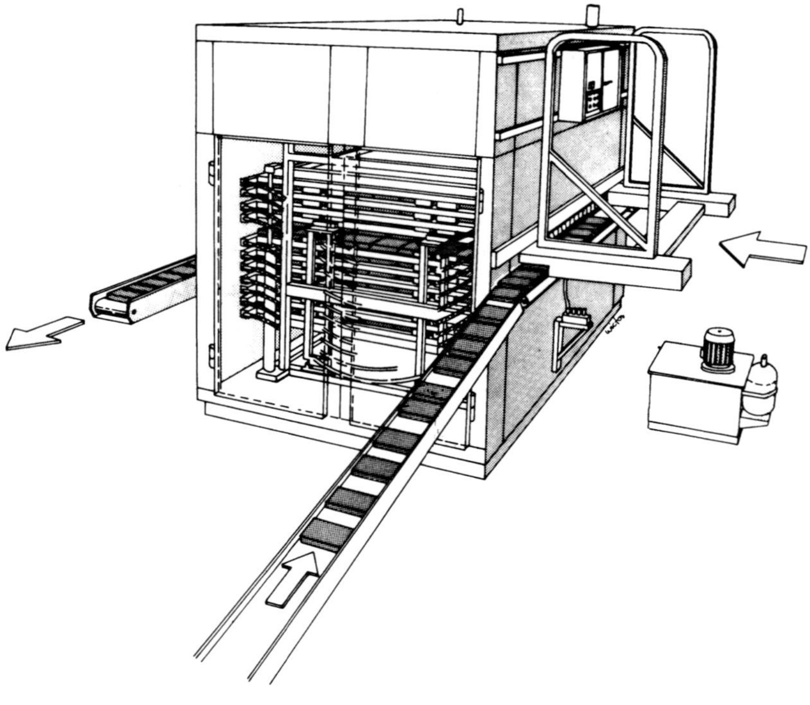

Figure 5.7 A semi-automatic plate freezer (courtesy Frigoscandia Contracting AB).

5.4.3 *Continuous freezers*

Continuous freezers are of more use in large-volume production. A steady feed of product can be channelled through the freezer. Because transport in to and out of the freezer is continuous, the process is less labour intensive than the batch freezer because there is no need to load and unload carrier racks, as required in batch freezing. An additional advantage of continuous freezing is that more uniform process conditions can be obtained, because the configuration of product in the freezing zone stays unchanged. Continuous freezers can use a variety of transportation systems for moving the product through the freezer. Some of these are illustrated schematically in Figure 5.8. When dealing with unpackaged products, some form of a conveyor system can be employed. This conveyor can take several configurations. Access of the freezing medium to the product must not be adversely affected by the conveyor. In each case the transportation system has to be compatible with the method of heat transfer. For example, in an immersion freezer, if a belt is used, this belt must allow the product to be immersed in the coolant bath, and also must facilitate drainage for recovery of the immersion fluid.

Some examples of different types of continuous freezer are described briefly below:

5.4.3.1 *Continuous tunnel air-blast freezer.* In this freezer, air is circulated over the product at a temperature between $-30\,°C$ and $-40\,°C$, and with an air velocity of between 1.5 and 6.0 m/s. The high air velocity gives good heat transfer. A conveyor belt carries the product through the insulated tunnel. In many operations, a series of conveyor belts are used, with product falling from one belt to another, thus breaking up clumps. The air flow may be either perpendicular to, or parallel to the vegetable flow. Moisture transfer into the air flow may result in a build up of ice on the cooling coils. Typical production rates are 200–1200 kg/h. The tunnel can also be used with cryogenic cooling. In this case cryogen is delivered rather than chilled air. The product still moves on conveyors, and sprays and fans introduce and move the cryogen within the tunnel.

5.4.3.2 *Spiral belt air-blast freezer.* These freezers are modifications of the continuous tunnel freezers. The conveyor is a flexible mesh which is formed into spiral tiers. The belt may run on support rails, or it may be self-stacking (Figure 5.9). In most cases the cold air is directed down through the belt stack, while the belt lifts the product. Thus there is a counterflow of air and product. Air temperatures and velocities are similar to those employed in continuous tunnels. The spiral belt freezer requires less floor space than the continuous tunnel. Some are capable of product throughput of up to 3000 kg/h. Once again, this freezer design is also suited for the use of cryogenic refrigerant.

5.4.3.3 *Fluidised bed air-blast freezer.* The fluidised bed air-blast freezer does

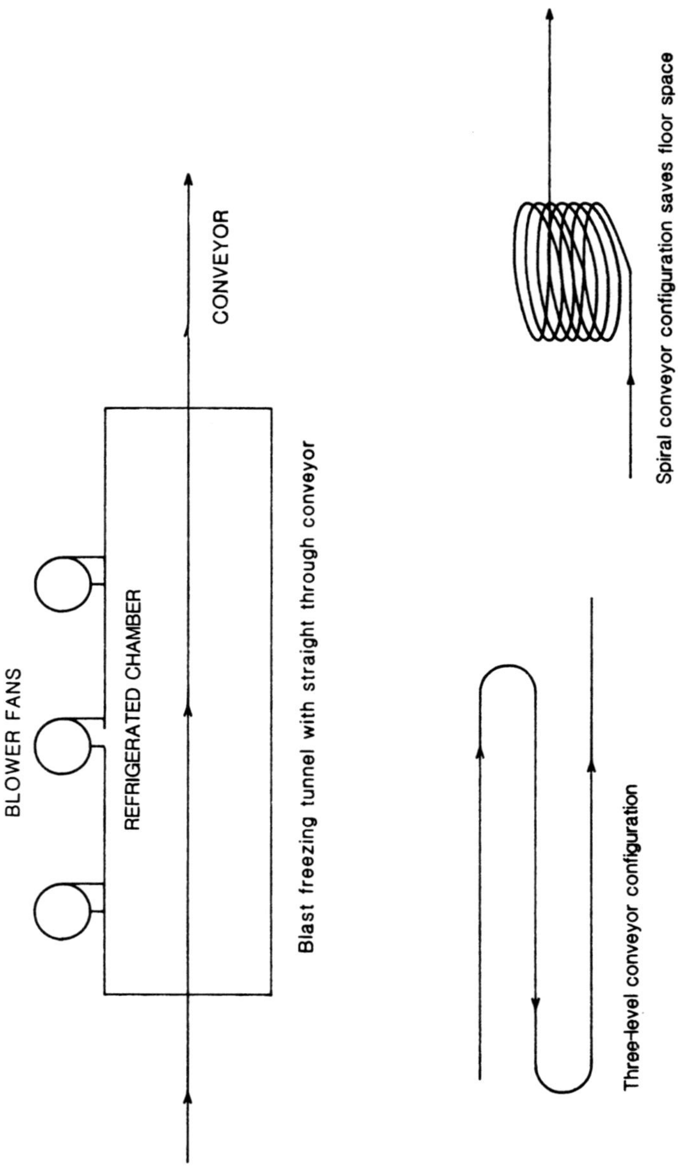

Figure 5.8 Schematic illustration of continuous methods of freezing.

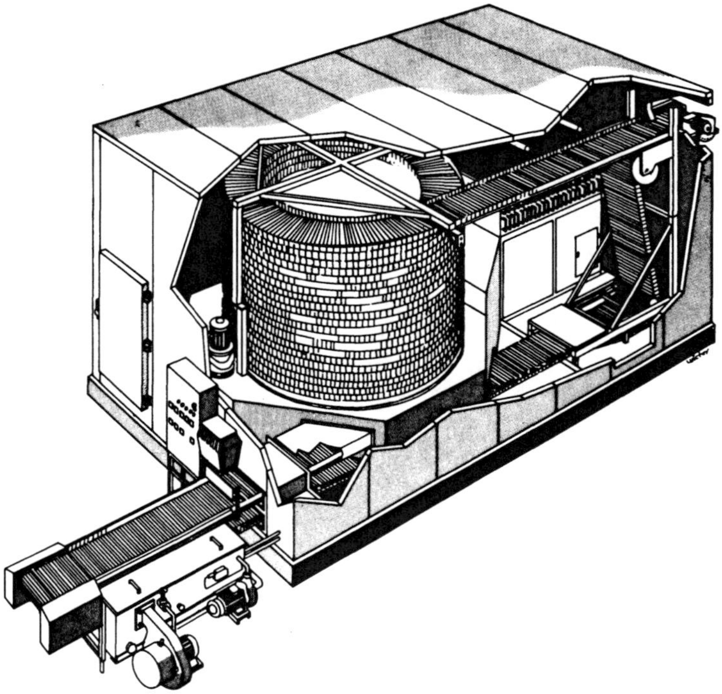

Figure 5.9 A spiral belt air-blast freezer – Gyro Compact (courtesy Frigoscandia Contracting AB).

not employ a conveyor (Figure 5.10). Instead, the fluidisation principle is employed. Air between −30 °C and −40 °C is passed at high velocity (around 5 m/s) up through a 3–14 cm bed of small vegetable pieces contained in a trough with perforated base. The vegetable behaves like a liquid, with turbulent movement. Heat transfer is very efficient. Flow control is by weirs. The thickness of the bed, and the air velocity required for fluidisation, depend upon product size and shape. Peas and diced carrots are ideal candidates for fluidisation. Moisture losses are reduced. Production rates of 10 000 kg/h have been reported.

As indicated, each freezer described has different performance characteristics. The choice of the most appropriate freezer has to be made with proper regard for the characteristics of the product to be frozen. Products rapidly frozen as individual pieces in a belt, spiral or fluidised bed freezer are often referred to as IQF (individually quick frozen) products.

5.5 Product characteristics

The following section gives brief details of the freezing characteristics of some typical vegetables, and of the methods most commonly used for their commercial processing.

Figure 5.10 A fluidised bed air-blast freezer – Flo-Freeze (courtesy Frigoscandia Contracting AB).

5.5.1 *Peas*

Peas are harvested at early mature stage. The maturity is often assessed by using a simple instrument such as a 'Tenderometer'. After shelling, peas may be separated by a brine flotation method, which separates fractions of different maturity. Blanching is required for a stable product. Peroxidase is used as indicator enzyme, though recently it has been suggested that blanching to a lipoxygenase-negative endpoint would give a better product. Because peas are small, rapid heat exchange is practical. Freezers employed include fluidised bed freezers, which are well suited to pea freezing, and continuous belt tunnel freezers both of which give IQF peas, which can be packaged immediately, or placed in bulk storage. An alternative, non-IQF method uses plate freezers. In this case the peas are pre-packaged. Heat transfer is not very efficient. There is also a tendency for moisture migration within the package. The more rapid freezing methods tend to produce a better quality product.

5.5.2 *Green beans*

Green beans after harvest are cleaned at the plant using shaker screens with blowers, followed by washing. Size grading, using specialised machinery, is carried out. Depending upon pod size, beans may be frozen whole, or pods may be cut. The size of material to be frozen ranges from 2 to 4 cm. Blanching is performed at 99 °C and usually only 2–3 min are required with peroxidase used as indicator enzyme. Once more, a less severe (90 s or less) blanch to a lipoxygenase negative end point may result in an improved product. Freezing is by continuous tunnel belt freezer for an IQF product, which may be packaged immediately, or bulk stored for later packaging. Pre-packaged material can be frozen, but the product is inferior. Green beans are particularly sensitive to temperature abuse in storage.

5.5.3 *Carrots*

Carrots may be frozen whole, sliced or as diced carrots. Harvested carrots must be washed to remove field dirt. Roots are then sorted into size classes, small, medium and large, which may be used to produce whole, sliced and diced carrots respectively. The carrots pass through cutters which trim off the ends, and are then peeled, in either a steam peeler or an abrasion peeler. Larger carrots are then sliced or diced, prior to balancing. Blanch times range from 2 to 8 min depending on product size. Diced or sliced carrots may be frozen in a fluidised bed freezer, or in a belt freezer. Small whole carrots are frozen on a belt freezer.

5.5.4 *Broccoli*

Broccoli deteriorates rapidly, so it must be cooled soon after harvest. It is also relatively fragile and merits careful handling. Broccoli passes into the plant on a conveyor line where substandard material can be removed, and the broccoli heads can be trimmed to more uniform size. Excess material trimmed off is used for chopped broccoli. Stalks are trimmed to around 12 cm lengths, and spilt lengthwise into uniform units. These are washed and blanched for around 3 min at 98 °C, the exact temperature and time depending upon product characteristics. Freezing should be rapid, so continuous belt tunnels are used. The material is too large, and too fragile, for fluidised bed freezing. Cut broccoli may be pre-packaged and frozen in a plate freezer or blast freezer.

5.5.5 *Cauliflower*

Cauliflower is handled in a manner very similar to that used for broccoli. Rapid cooling is required. The heads have their base removed, and then the curd is broken up into florets by one of a variety of means. Florets larger than about 5.5 cm are divided. Grading removes small pieces. After washing, the florets are steam blanched for about 4 min, and rapidly cooled. Florets may then be frozen by the IQF system on a belt freezer, or be pre-packaged and frozen in a plate freezer or blast freezer. Frozen product may be placed in retail or food service packs immediately, or may be bulk stored for later packaging.

5.5.6 *Brussels sprouts*

Brussels sprouts are size sorted, and small buds removed. They are then trimmed to remove the butt end, and the outer leaves. Small, medium and large sprouts should be processed separately, because of different processing requirements governed by the different heat exchange characteristics. Medium-sized sprouts need about 5 min blanch time in steam. Sprouts may be frozen by IQF methods, or packaged prior to freezing. If pre-packaged, plate or blast freezers may be used.

5.5.7 *Spinach*

Spinach is cut before the leaf becomes too long, and before it starts to go to seed. Because the leaves tend to form a tightly packed mass which is difficult to cool, it is necessary to minimise heat pick up. Air cleaning on a mesh belt is often employed, followed by inspection and washing in a turbulent tank, spray washing on a belt, with subsequent drainage on a mesh belt. All traces of field soil must be removed. Blanching is in steam or boiling water for about 2 min. This makes the leaves limp, and capable of forming an impenetrable mass. Care must be taken not to overload the blancher. Immediate cooling follows the blanch. A mesh belt is used for drainage. The spinach is normally packaged before freezing. Any kind of freezer is suitable.

5.5.8 *Sweet corn*

Sweet corn may be frozen as whole kernel corn, or as corn-on-the-cob. Timing of the harvest is important as the kernels must still have a high sugar content. Freezing must be done within hours of harvest. Corn-on-the-cob presents many problems, because of its large size. Blanching and cooling are difficult. A blanch time sufficient to inactivate enzymes at the centre of the cob will result in a too severe treatment of the kernels. Many blanch conditions are used and times are usually between 8 and 12 min in steam. Whole kernel corn may be blanched in a number of ways; it may be blanched on the cob before cutting, a two-stage blanch may be used (before and after cutting), or blanching may follow cutting. The latter method will result in higher product loss, but lower energy cost, because the cob does not need to be heated. The two-stage blanch is often considered a good compromise. The cut corn must be cleaned to remove debris. Whole kernel corn is frozen by IQF methods. It may then be packaged, or bulk stored. Corn-on-the-cob is often belt frozen.

5.5.9 *Potatoes*

Potatoes are a major frozen product, and may be frozen in a variety of forms. Tubers may be stored for a considerable time before processing. Before freezing, the potatoes are washed and peeled. A variety of peeling methods are available, including high-temperature lye peeling, abrasive peeling, and steam peeling. Peeled potatoes are trimmed, inspected and size sorted. Small potatoes are often used to produce mashed potatoes, croquettes and other similar products. The larger potatoes are sliced to produce long cuts for frozen chips (American French fried potatoes). A variety of sliced, fried products are possible. The potato strips are water blanched, often at 90 °C for about 10 min, to modify the surface, rather than to inactivate enzymes, before frying. The product is then fried, the excess fat is removed, and freezing takes place on a continuous belt freezer, at around −40 °C. This takes around 12 min. Pre-packaged product takes around 3 hours to freeze in a plate freezer.

Potatoes are an example of a vegetable which is processed to a form close to its final form prior to freezing. Most vegetables are frozen in a form much closer to that of the original raw material, leaving the end user to perform key preparative steps. Many modern operations take the frozen vegetable and use it as a component in a value-added product, such as vegetable dishes in sauces, mixed vegetable preparations, and complex dishes, and whole meals. It is generally true that the vegetables used for these purposes have previously been frozen as a bulk commodity, and the new product is a repackaging operation in terms of the vegetable component.

5.6 Storage

One final aspect of the freezing of vegetables that must be considered is that of frozen storage. It is generally true that storage life is greater at lower storage temperatures, and temperatures of −20 °C or below are often recommended for the long-term storage of frozen vegetables. Temperature fluctuations during storage can reduce this storage life. While it has been shown for many systems that the effect of temperature fluctuation is similar to maintaining storage at the average temperature of the fluctuation (16, 17), this is not always the case. Physical processes are often accelerated by fluctuating temperature, especially those processes which involve the migration of water. This can be seen, for example, in moisture migration in packages. Temperature fluctuations lead to a rapid build up of water on the internal surfaces of the package. In some cases freezer burn, which is a combination of surface dehydration, followed by surface oxidation, can be a problem. Temperature fluctuations can also accelerate the growth of ice crystals within the product. For green beans, even a relatively short exposure to higher subfreezing temperatures can lead to detectable colour differences (as compared to a well-maintained control) after a few additional months of low-temperature storage. This abuse enhanced colour degradation may occur in other vegetables too. It is therefore advisable to maintain properly controlled low-temperature storage, and to guard against temperature fluctuations. Packaging, too, will affect storage stability. Moisture permeable packages will result in dehydrated products. Access of oxygen should be minimised by the package to help control oxidative degradation. The vapour spaces in the package should be held to a minimum, as otherwise conditions are ideal for internal moisture migration.

References

1. Fennema, O.R., in *Low Temperature Preservation of Foods and Living Matter* (eds Fennema, O.R., Powrie, W.D. and Marth, E.H.), Dekker, New York, 1973, Chap. 4.
2. Brown, M.S., *Adv. Food Res.* **25** (1979), 181.
3. Reid, D.S., *Int. J. Refrig.* **3** (1980), 226.
4. Heldman, D.R. and Singh, R.P., *Food Process Engng*, AVI, Westport, CT, 1981.
5. Reid, D.S., *Food Technol.* **37** (4) (1983), 110.
6. Reid, D.S., in *The Effects of Low Temperatures on Biological Systems* (ed. Grout, B.W.W. and Morris, G.J.), Arnold London, 1987, Chap. 15.
7. Powrie, W.D., in *Low Temperature Preservation of Food and Living Matter* (eds Fennema, O.R., Powrie, W.D. and Marth E.H.), Dekker, New York, 1973, Chap. 7.
8. Joslyn, M.A., *Adv. Enzymol.* **9** (1949), 613.
9. Lund, D.B., in *Principles of Food Science, Part II: Physical Principles of Food Preservation* (eds. Karel, M., Fennema, O.R. and Lund, D.B.), Dekker, New York, 1975, Chap. 3.
10. Poulsen, K.P., *Food Technol.* **40** (6) (1986), 122.
11. Potter, N.N., in *Food Science*, AVI Westport, CT, 1986, Chap. 18.
12. *Commodity Storage Manual*, The Refrigeration Research Foundation, Washington, DC (see listing for each vegetable).
13. Bottcher, H., *Nahrung* **19** (1975), 173.
14. Williams, D.C., Lim, M.H., Chen, A.O., Pangborn, R.M. and Whitaker, J.R., *Food Technol.* **40** (6) (1986), 130.
15. Hallowell, E.R., in *Cold and Freezer Storage Manual*, AVI Westport, CT, 1980, Chap. 2.
16. Olson, R.L. and Deitrich, W.F., in *Quality and Stability of Frozen Foods* (eds van Arsdel, W.B., Copley, M.J. and Olson, R.L.), Wiley, New York, 1969, Chap. 5.
17. Moleeratanond, W., Kramer, A., Ashby, B.H., Bailey, W.A. and Bennett, A.H., *ASHRAE Trans.* **85** (2) (1979), 56.
18. Williams, M.L., Landel, R.F. and Ferry, J.D., *J. Amer. Chem. Soc.* **77** (1955), 3701.

6 Chilling

CHRIS BRIMELOW and DHARAM VADEHRA

6.1 Introduction

Sales of value-added chilled foods are on the increase in many of the more sophisticated countries of the world. Usually, the chilled food sector in these countries is the fastest growing area of food retailing. Chilled foods are also playing an increased role in catering and food service applications. In the European food market, chilled food sales account for approximately 43% of the total food market volume (1988 figures). This proportion can be compared to 4% for frozen foods and 22% for bottled food and drink. In the USA, the proportion is slightly lower, due to the very high penetration of frozen foods in the US marketplace.

Much of the volume in the chilled food market is made up of commodity and staple items (fresh meat, milk, cheese, butter), and traditional items (charcuterie, bacon, sausages, baked products). However, the positive growth in the chilled product sector comes either from new food groups, such as prepared meals and salads, pizzas, pasta products, desserts and yoghurts, or from value-added versions of the old commodity items such as pre-packed prepared vegetables, prepared cuts of meat and pre-packed fish.

The trends in the sales of chilled foods are closely linked with other consumer and market trends which are very evident and have been discussed in many marketing publications. Some of these changes are: more working women/dual income households, more single-person households, decline of the youth segment, larger affluent middle-age group, better informed consumers, increase of leisure time and time awareness, changes in shopping habits (one-stop shopping), increase of refrigerator and microwave oven penetration, polarisation of food habits (functional: enjoyable), fragmentation of meal patterns, increase in health awareness, increased interest in ethnic foods and changes in packaging preferences.

Two or three of these trends in particular have stimulated consumer interest in chilled foods: mention can be made of the awareness of health issues, the move towards foods having a more natural appearance and away from foods having a processed overpackaged appearance and the increased desire for convenience as exemplified by the penetration of refrigerators and microwave ovens. A recent Delphi study carried out for the EEC by Young (1) concluded that sales of chilled foods in Europe would double by the year 2000, whereas frozen food sales would increase by 37% and canned food sales would decrease by 30%.

One group of products which demonstrates a number of the perceived advantages of chilled foods is the group constituted by vegetable products and products containing vegetable components. Firstly, chilled vegetables can meet the sensory criteria of natural appearance, attractive flavour and desirable texture required by the consumer, whereas vegetables from other technologies, such as frozen, canned or dehydrated, often fail to meet one or more of these criteria. Secondly, for the health conscious consumer, chilled vegetables provide a source of water-soluble vitamins and carotenoids, and also a source of fibre. This is especially relevant at this time, when concerns over the lack of fibre and the excessive amounts of fat in the modern diet have been voiced by so many expert nutritionists. Thirdly, improvements in chilled food processing technology have led to more convenient vegetable products, pre-washed, peeled, pre-cut, portion controlled and pre-packed, with extended shelf-lives compared to the shelf-lives achievable a few years ago.

In short, chilled vegetable products are perceived to offer a desirable combination of a natural wholesome appearance, flavour and texture with good nutritional properties and ease of preparation. The word 'perceived' is used advisedly: not in all cases are perceptions borne out by facts.

This chapter will firstly describe the processing method available to meet the challenges and opportunities in the chilled vegetable product marketplace, and will then go on to discuss some of the effects of these processes on quality and shelf-life.

6.2 Historical and commercial aspects of chilled vegetables

One of the earliest applications of food preservation technology was in the area of vegetables, because seasonal factors meant that fresh vegetables were only available for a short period of time. Canning, pickling and dehydration as applied to vegetables are described elsewhere in this book. More recently, a number of factors have made the need for vegetable preservation less relevant. Firstly, agronomists have met with much success in extending the growing and harvesting season for many vegetables. Secondly, improved methods of transportation, the removal of international trade barriers, improvements in national food distribution systems and improvements in stock control procedures through distribution and retailing, have all contributed to a more efficient supply chain, and hence to a reduction in the need for long shelf-lives. Thirdly, as will be described, improvements in refrigeration and chilling processing technology have also led to extended shelf-life chilled vegetables. Because, in this chapter, much reference will be made to extension of shelf-life, it is wise at this juncture to give a definition of the term. Shelf-life has been well defined by Shewfelt (2) as the time period that a product can be expected to maintain a predetermined level of quality under specified storage conditions.

The historical origins of the application of cold as a means of extending the shelf-life of such foods as commodity vegetable produce, are shrouded in the mists of time. It has been common knowledge for centuries that reduced storage temperatures will retard or even arrest the deterioration of many fresh foods. In some countries, ice blocks were delivered from colder regions to the homes of the wealthy for such a purpose. The European cellar store was also a means of maintaining a reduced temperature.

The invention of compressed air and compressed ammonia refrigeration systems at the end of the last century was the next major step. Some of the original developments in this field have been described by Rolfe (3) in an excellent early review of the subject of chilling and freezing of foodstuffs. Home refrigerators and retail outlet refrigerated display counters came into more common use in the twenties and thirties. The consumer quickly became familiar with bulk refrigerated displays of commodity vegetables. However, little effort was made by the suppliers of such produce to provide added value to the consumer beyond the use of simple techniques such as washing, size grading, bagging and bundling.

A more recent trend, which has occurred over the last ten to fifteen years in the marketplaces of the more sophisticated countries of the world, is towards raw, minimally processed produce. With this type of product, some degree of added value is provided to the consumer or the food service outlet by, for example, pre-preparation for convenient use (peeling, dicing, cutting or slicing), by providing the product in attractive and convenient protective packaging and by combining several vegetable items in the same package.

A third type of chilled vegetable product, marginally heat-processed vegetables, has appeared in the marketplace much more recently. The concept, which was pioneered in the UK in the late 1970s, has spread to other countries in Europe and to Japan and, even more recently, to the USA. These products provide a high degree of added value to the consumer. The vegetables are fully prepared and are usually par-cooked or pasteurised. They are available particularly as components in ready-to-eat or ready-to-heat prepared meals or salads.

6.3 Process treatments for minimally processed vegetables

A general review on the subject of minimally processed vegetables has been given by Huxsoll and Bolin (4). Shewfelt (5) has defined minimal processing as 'handling, preparation, packaging, and distribution of agricultural commodities in a fresh-like state'. This includes any process that adds value, but short of conventional food preservation techniques such as heat preservation, freezing and dehydration. In view of the great effect which a heat process can have on the microbial and sensory quality throughout the shelf-life, minimally processed vegetables can be divided into two types: (A) those which undergo no heat treatment, and (B) those which are heat treated. In type

A products, the vegetables are in the raw state and are thus capable of undergoing all of the storage changes associated with respiring fresh materials. These changes include vitamin losses and other physiological changes, all of which are aspects concerned with quality. Some of these changes can be more or less slowed down and shelf-life, therefore, extended by the choice of suitable processing conditions. In type B products, the heat process is usually sufficient to inactivate respiration associated enzymes and other factors come into play to determine the shelf-life.

In both types, the desired quality and safety criteria can only be achieved throughout the shelf-life by utilising a combination of treatments, any of which if individually applied would not result in achievement of all the criteria. These treatments have sometimes been termed hurdles, after Leistner and Rodel (6). Treatments are chosen such that, firstly, they do not cause the product to lose its essential fresh-like character, and, secondly, they are not mutually incompatible. Additive or synergistic effects between treatments are often encountered and can be exploited.

Some of the individual treatments which may be part of a combination process include:

- Low-temperature storage
- pH modification
- Reduction of water activity
- Surface coatings
- Chemical dips and preservatives
- Addition of competitive microflora
- Mild heat treatment (type B products only)
- Sanitary packaging
- Controlled atmosphere (CA) or modified atmosphere (MA) storage
- Speed of chilling
- Irradiation.

In view of regulatory requirements or reasons associated with the perception of freshness or wholesomeness, it is often not possible to use all of these treatment together. Thus, the industry is restricted, normally, by not being able to use irradiation and by the limited applicability of such treatments as pH reduction, water activity reduction, use of competitive microflora and the addition of chemical preservatives or dips.

6.3.1 *Storage temperature*

The overriding factor controlling shelf-life with all of these treatments is the storage temperature itself. As predicted by the van't Hoff rule, Q_{10} values for vegetables (that is, the increase in rate of a deteriorative reaction for each 10 °C rise in temperature) are in the order of 2–3 in the 0 to 25 °C temperature range (7). Thus, many vegetables which will remain at good quality for two weeks at 1 °C will only last 3–4 days at normal ambient temperatures.

Though classical kinetics are followed over most of the temperature range, the so-called temperature anomaly occurs at or near the freezing point of vegetables and other foods. Thus, broccoli stored at its freezing point (−1.7 °C) has twice the shelf-life of broccoli stored at 0 °C. The temperature anomaly is one of the reasons behind the development of superchill technology in Japan (known as *hyo-on* in that country) (8). In this technology, food material is chilled to very close to its freezing point, usually between −3 and 0 °C, and is then maintained at this temperature throughout the chill chain, from the manufacturer's chill store through distribution and retailing to the consumer's refrigerator. Refrigerators are equipped with special compartments to store superchill items. Benefits are claimed for this system in terms of both microbial and sensory quality. Some European manufacturers of domestic and retail refrigerators are actively examining the business opportunities of superchill in the European market.

However, the benefits of superchill with regard to many raw vegetables are not always obvious in view of the sensitivity of vegetables to chilling injury (CI) or cold-shock. The subject of CI has been extensively reviewed recently by Wang (9) and by Jackman *et al.* (10). It is well known that most crops of tropical and subtropical origin, and many of temperate zone origin, are susceptible to CI. CI can be enhanced by preparation treatment prior to chilling.

CI may take the form of discoloration, textural changes towards limpness or rubberiness, or surface changes such as pitting. Some methods are available to alleviate the development of CI symptoms; these include pre-conditioning at higher chill temperatures, modified atmosphere storage or chemical treatments, for example, SO_2, and calcium dips. Both reviews mentioned include tables of critical CI temperatures for different vegetables.

Despite problems with CI, in general terms chilling of vegetables extends their shelf-lives. Temperature control throughout the chill chain from manufacturer's warehouse through distribution and retail to the consumer's refrigerator is, therefore, of key importance (q.v.). Application of other process technologies to extend refrigerated shelf-life will be irrelevant if temperature control is not practised. Reduction of temperature restricts microbial growth; however, minimally processed vegetables will still be subject to cold-tolerant microorganisms (psychrotrophs) unless further steps are taken to eliminate or minimise the risks from these organisms (11). Some of the steps available to the chilled vegetable processor will be discussed below.

6.3.2 *pH modification*

Lowering of the pH can significantly affect the shelf-life of chilled vegetables; however, it has limitations in that the organoleptic changes produced may be unacceptable for some products. One problem, for example, is that reduction in pH causes green vegetables significantly to lose their green pigment, turning olive green due to the conversion of chlorophyll to pheophytin. However, in combination with mild heat treatment, reduction of pH can provide some significant advantages.

6.3.3 *Water activity*

It is well known that the ability of different microorganisms to grow in a food is related to the water activity of the food. Water activity (a_w) can be defined as the ratio of water vapour pressure of the food system (VP) to the vapour pressure of pure water at the same temperature (VP_w), that is,

$$a_w = VP/VP_w$$

The a_w of a food can be reduced either by adding solutes (humectants) to the water phase or by reducing the concentration of water phase. Both these methods have the effect of limiting the water available for microbial growth, thus retarding or even preventing such growth. Different interactive effects between a_w and such extrinsic variables as storage temperatures and atmosphere composition have been observed.

Data on the natural water activity of different vegetables and other materials has been given by Chirife and Ferro Fontan (12). Normally, the water activities are between 0.970 and 0.996. It is apparent that to gain an effect on microbial stability by reduction of water activity, the organoleptic properties of the vegetable material would be changed either due to a reduction in water content or by the incorporation of humectants or both (4). These changes would not normally be acceptable for a chilled vegetable product having fresh, natural appeal. An examination of the sparse literature relating to the use of reduction of water activity in the production of chilled vegetable products bears this statement out. Deak (13) described laboratory experiments in which the growth characteristics of different microorganisms isolates from stored plant products were examined under different chill temperatures (5–15°C), different water activities (0.92–1.0) and different atmospheres. Results indicated that increased CO_2 and decreased water activity and temperature gave the greatest inhibition of microbial growth as expected, but that when test vegetables were studied (cauliflower, paprika) these optimum conditions did not necessarily correspond to optimum conditions for storage of the vegetables. Reduction of water activity may, however, be applicable to such products as chilled pastas with vegetables fillings (ravioli, tortelloni), where fresh appearance and fresh texture properties in the vegetable component are not key to the success of the product.

6.3.4 *Surface coating systems*

Another method of extending the shelf-life of raw minimally processed vegetables is the use of edible coating technology. Edible coatings can function in a number of ways: to retard loss of moisture from vegetable pieces by offering a physical barrier to migration; to retard oxygen uptake by the vegetable matter thus slowing down oxidative deterioration reactions; to offer a surface preservative or antimicrobial action. Kester and Fennema (14) have reviewed the recent literature relating to edible films and coatings. They have divided films into four types: polysaccharide films, such as alginate, pectin, starch, carrageenan and cellulose derivatives, which all have limited moisture barrier properties but may have some resistance to oxygen permeability; protein films such as gelatin, casein and serum albumen, which again have limited moisture barrier properties; lipid films such as acetoglycerides, waxes and surfactants, which have much better moisture barrier properties than the first two types and offer some promise for use with vegetables; composite films which may be bi-layers of protein, lipid or polysaccharides. The latter type has the greatest applicabiliy for use on vegetables. Different materials can be incorporated into edible films and of particular relevance to chilled vegetables are preservatives and antioxidants.

Wax coatings have traditionally been applied to fresh vegetables to extend post-harvest storage life by limiting water loss and slowing down respiration. Vegetables which have been wax coated include turnips, carrots, sweet potatoes, cucumbers, rutabagas, squash, potatoes, tomatoes and peppers (15, 16). Jojoba oil has occasionally been used as a replacement for the beeswax or carnauba wax normally employed (17).

Composite films are also frequently used on fresh vegetables. Examples are CMC or sodium CMC plus fatty acid esters of sucrose which have been utilised on many materials (18), a carboxylic acid salt plus an alkylamine utilised by Everest-Todd (19) for potatoes, red beet and carrot, and a novel film system developed by Kirsop and Brocklehurst (20) for cut salad vegetables which are added to mayonnaise to make coleslaw. This latter system is produced by a three-stage dipping process consisting of a first-stage dip solution comprising propylene glycol alginate, calcium lactate, a surfactant and an antimicrobial agent, a second-stage solution of alginic acid and a third-stage solution of calcium lactate. This treatment is claimed to extend shelf-life and eliminate the aged or translucent appearance of the vegetables in the coleslaw. The use of edible films for extending shelf-life of value-added products such as delicatessen salad mixes has great potential for the future.

6.3.5 *Surface dipping systems*

A method to extend the shelf-life of raw or cooked prepared vegetables is the use of surface dip systems. Surface dips can have a preservative action, can delay sprouting or can have other beneficial effects in terms of sensory attributes such as reducing surface browning or maintaining firmness.

Traditionally used preservative dips include organic or inorganic acids, chlorine and sulphur dioxide. Sulphur dioxide functions both as a preservative and as an anti-browning agent. The use of sulphur dioxide for raw, peeled, chilled vegetables has been reviewed by Baardsath (21). The recent controversy surrounding allergenic reactions associated with sulphur dioxide on salad vegetables has prompted the investigation of alternatives. Rice (22) has examined mixtures of ascorbic and citric acids, mainly from the point of view of browning inhibition. Such combinations preserved sliced iceberg lettuce in permeable polyethylene bags for 5 days at 38 °F compared to 7 days for sulphur dioxide. Miller *et al.* (23) investigated chlorine and imazalil alone or in combination as preservative dips for green bell peppers. Kratky and Vadehra (24) have described the use of a dipping stage involving solutions of different antimycotic agents such as hydrogen peroxide, chlorine, sorbic acid or benzoic acid, before a heat-processing stage (50–70 °C) as a method of improving shelf-life, as well as the colour, of green vegetables stored at 4–8 °C. Goussalt (25) describes a method of extending the shelf-life of vegetables by dipping them in weak vinegar, partially drying, spraying with ascorbic acid, chilling and packing in a film permeable to water vapour.

Sprouting inhibition has been examined recently by a number of workers (for example, Passam (26) used gibberellic acid dips for yam tubers; Malik *et al.* (27) used corn oil dips for potatoes), but none of these studies appears to relate to chilled vegetables.

Extension of sensory shelf-lives by means of surface dips has taken many forms. Calcium chloride dips have been widely used to improve the firmness of fresh vegetables, (for example, in asparagus (28)). Surface browning of fresh and chilled vegetables has been inhibited by the use of ascorbic acid (for example, in okra (29)), aqueous emulsions containing oleic acid (for example, for various vegetables (30)) or sulphur dioxide (for example, for potato chips (31)). Benzyladenine dips have been employed to improve the colour, texture, aroma and flavour of raw and cooked broccoli stored at 2 and 5 °C (32).

6.3.6 *Addition of competitive microflora*

The use of competitive microflora to extend the shelf-life of chilled vegetable products has a very limited application and will not be covered at any length. The most common of the microbial methods is vegetable fermentation using lactic acid bacteria. This results in a mildly to strongly acid material which can be added to salad vegetable mixes in certain cases.

6.3.7 *Heat treatment*

The primary purposes of a heat treatment are to decrease the initial microbial load and inactivate enzymes. In some cases a heat treatment also improves the colour and flavour. The time and temperature conditions of the heat process are very critical as they will not only affect the sensory quality, but also the shelf-life.

For products where the crunchy texture of the raw vegetable is a desirable feature, such as in salads, the heat treatment should ideally aim at not disturbing the vegetable membranes excessively while at the same time inactivating the enzymes which cause deterioration. Enzymes and substrates are generally compartmentalised in vegetables and several of the enzymes are bound to the membrane. Kratky and Vadehra (24) have demonstrated that at temperatures at or just below 60 °C, a process of around 3 min can decrease the initial vegetative microbial load significantly without affecting the texture and colour of green vegetables detrimentally. The product generally remains green through a 3–4 week shelf-life and the texture remains firm and crisp. These mild temperatures do not damage the membranes and the behaviour of the vegetables is more like fresh than processed vegetables. Mildly heat-treated vegetables can still contain heat-restraint vegetative cells and surviving spores which may grow during chill storage. Therefore, either superchill storage conditions should be used or other combination treatments may be needed to ensure the required safety and shelf-life.

For products where raw vegetable crunch is not a requirement, standard blanching pasteurisation conditions can be carried out. Pasteurisation values ('P' values) can be defined by reference to the effect at a given temperature, thus a process having a P_{85} value of unity is equivalent to 1 min at 86 °C for an organism having a z value of 10 °C, assuming instantaneous heating and cooling (33). In work carried out at various Nestlé R and D laboratories on different vacuum-packed ready meals, including some with vegetable components, it was found that in order to achieve repeatable results, particularly with reference to microbial shelf-life, it is necessary not only to specify the centre point pasteurisation value, but also the process temperature (that is, the retort operating temperature). This is because the microbial quality is also affected by the temperature gradient. Assuming well-prepared raw materials, a shelf-life of 21 days or more at 8 °C is usually achievable with in-pack treated vacuum-packed vegetable products if a P_{85} process of 200 is given in a pasteuriser run at a water temperature of 100 °C. As a comparison, if the pasteuriser is run at a water temperature of 116 °C, the same products could be processed to P_{85} values of 60 and would achieve a similar or longer shelf-life.

6.3.8 *Speed of chilling*

The chill-down time and temperature parameters in a chilled food process is one of the critical control points in the assurance of safety (34). The UK Department of Health and Social Security (35) has issued guidelines for the production of pre-cooked chilled foods in which it is recommended that reduction of the core temperature to below 3 °C should occur within 120 min of the end of the cook process and that the chilling process should commence not later than 30 min after the end of cooking.

There are a number of reasons for carrying out rapid chilling. Firstly, it reduces the risks of growth of surviving microbes occurring in the critical warm zone between 50 and 10 °C. Secondly, it ensures that only pre-chilled product enters storerooms or distribution vehicles, where the refrigeration units are designed to hold chill temperatures, not to cool down the product. Thirdly, slow cooling can result in flavour and texture losses and other quality deterioration, compared to rapidly cooled product. Claims have been made for a synergistic effect on microbial inhibition, between the heat process and the rapidity of chilling, for some systems ('chill shock').

6.3.9 *Sanitary processing and packaging*

The important difference between chill food technology and other technologies with respect to processing requirements is in the area of hygiene. Hygiene levels in the manufacture of chill foods generally have to be superior to those applicable to other technologies and this applies to machine hygiene, factory fabric hygiene (floors, walls, ceilings, fitments), sanitary practices by operators and factory air quality.

In order to minimise cross-contamination and to isolate zones of different hygiene levels, chill foods factories are usually laid out and organised by what has been termed the fortress or isolation principle. This technique was pioneered for manufacture of chill foods by Marks and Spencer in the UK in the 1970s (36) but it now common practice in the industry. High-risk zones are isolated from lower risk zones, not only by walls and ceilings, but also by separating the air-handling systems and by controlling and sometimes preventing the movement of operators between the zones. Air overpressures in high-risk areas are arranged to be higher than those in neighbouring lower risk areas. Access to areas of higher risk are often through pressure locks. So, for example, it is not possible to open the door into the higher risk zone without previously closing the door from the lower risk zone.

The packaging area is normally the zone of highest risk. For some particularly sensitive product ranges, such as those where components are assembled after cooking or where components do not lend themselves to adequate pasteurisation, it may be advisable to utilise full clean-room systems with highly filtered air (Class 100) and clean-room garbed operators (37). This technique has been applied by a number of manufacturers, including the Nestlé Company for ready meal ranges manufactured in the UK, Sweden and the USA.

The designation of low-, medium- and high-risk zones will differ from product range to product range, but with vegetable products it is normal to separate raw uncleaned vegetable receipt and storage areas from cleaning and preparing areas, prepared vegetables areas from cooking areas, cooking areas from packing areas, and packing areas from chilling and chill storage areas. If vegetables are to be one component in a product containing meat or fish and a sauce, it is also essential practice to separate vegetable storage and preparation zones from meat/fish storage and preparation zones and from sauce preparation zones. The product would then be assembled together either just before or after the cooking stage.

6.3.10 *Controlled atmosphere and modified atmosphere packaging*

Both of these techniques involve changing the atmosphere around the food material from air to another composition: usually the oxygen content is lowered and the CO_2 content is raised. Two methods are utilised, gas flushing and evacuation then regassing. With controlled atmosphere packaging (CAP) the composition of the atmosphere is controlled through the storage life by suitable choice of the packaging material permeability properties. In the case of modified atmosphere packaging (MAP), the atmosphere is changed at the packing point and no further efforts are made to control its composition; it is less often practised for prepared raw vegetables than CAP, but is the obvious technique for prepared marginally heat-processed vegetable products.

CAP technology is derived from the controlled atmosphere (CA) technology utilised for extending the life of bulk stored fruits and vegetables and widely practised in many countries. Gas-tight warehouse stores, which can contain many hundreds of tons of product, are equipped with systems which carefully control the gas atmospheres within the stores. CA technology for commodity products was originally developed in the UK in the 1920s and 1030s, particularly by Kidd and West (38, 39) for storage of apples and tomatoes. More recently, CA technology has been extended to shipping containers.

CAP technology was pioneered in the 1960s by Karel and his group at MIT in the USA (40, 41). Reviews on the subject have been written by Tomkins (42), Rizvi (43), Ben-Yehoshua (44), Labuza and Breene (45), Barmore (46) and Day (47). Despite being pioneered in the USA, CAP has found a wider practical application in the UK and Europe (48).

With CAP, the use of selectively permeable packaging films in association with a known packing gas composition provides an internal atmosphere of the desired composition during the shelf-life. In the sealed package, the oxygen level will fall and the CO_2 level will rise, due to natural respiration effects in the raw vegetable. If the package was totally impermeable, spoilage would occur fairly quickly due to anaerobic glycolysis at low oxygen pressures. A suitable semi-permeable packaging film allows ingress of oxygen at a controlled rate to replace the oxygen taken up by the fresh product. The lower the permeability of the film, the lower will be the final oxygen level. A steady state will be reached at a given temperature where the rate of uptake of oxygen by the product is the same as the rate of replacement from the outside atmosphere. This steady-state oxygen pressure value is dependent on such variables as the product, the film, the temperature and the gas composition of the internal and external atmospheres.

The steady-state value, therefore, can be arranged to be at the optimum for good storage of the product by suitable choice of the film and gas atmosphere. Oxygen permeation rates of packaging materials at ambient temperatures are readily obtainable from manufacturers though data at chill temperatures are more difficult to find. Some limited data are given in Table 6.1 which indicates that oxygen permeation at 2 °C/95% RH is between 2 times and 20 times less than that at 23 °C/55% RH. As has been pointed out by DeLassus and Hilker (49), permeability is an exponential function of temperature. A plot of log permeability (P) versus the reciprocal of the temperature ($1/T$) gives a straight line at a constant relative humidity, unless the glass transition temperature of the polymer (T_g) occurs within the temperature range concerned. In the latter case, the linear gradients above and below T_g would be different. The other factor affecting oxygen permeability rates is the water activity of the food sample: as the water activity rises the permeability of the film also rises. Optimal controlled atmosphere storage conditions for different fruits and vegetables, together with a comparison of the estimated shelf-lives, have been given by Labuza (50). Computer programs for calculation of steady-state conditions have been provided by Simon *et al.* (51).

Recent applications of CAP to fresh, minimally processed vegetables have been discussed by Ballantyne *et al.* for broccoli florets (52) and shredded lettuce (53), by McLachlan and Stark (54) for broccoli, Brussels sprouts, carrots, iceberg lettuce and potatoes, and by Geeson *et al.* (55) for tomatoes.

The use of MAP for the extension of shelf-life of marginally heat-processed vegetable products is a somewhat newer technique than the application of CAP to raw prepared vegetables. It has only been seen from the early 1980s, particularly in the UK and France. The technique is based on the use of nitrogen alone or in mixtures with carbon dioxide, and on the reduction of oxygen to levels normally less than 1%. Carbon dioxide is by no means inert and can affect the colour, texture and other quality attributes of cooked vegetables. The major advantage is related to the inhibition of the growth of moulds, and to some extent yeasts arising from post-process contamination. Bacterial growth is not as significantly affected by carbon dioxide.

More recent developments in MAP packaging have involved the use of gas scavenging films, gas scavenging sachets added to product packs, chemical preservative films and drying films (45, 46). Scavengers are available which reduce ethylene produced by respiring vegetables, reduce residual oxygen or reduce carbon dioxide. The usual oxygen scavengers are based on ferrous salts (56), sulphites (57), glucose oxidase and alcohol oxidase.

6.3.11 *Irradiation processing*

Irradiation as a preservation method or as simply a method of extending shelf-life of various foods has been extensively studied for over four decades. In 1981, the report of the Joint FAO/IAEA/WHO Expert Committee (58) concluded, after looking at the evidence, that there was no toxicological hazard associated

Table 6.1 Oxygen permeability data at chill temperatures.

Film	Supplier	Structure	Thickness (μm)	O_2 Permeability ($cc/m^2.d.atm$)	
				2 °C/95% RH	23 °C/55% RH
Campack	Camvac	12 μ met PET/50 μ LDPE	62	0.02	< 1
Hybar	BXL	LDPE/EVOH/LDPE	100	0.14	0.4
Sardnex XZ86513	DOW	LDPE/PVdC/LDPE	75	0.42	8
Bicor MB777	Mobil	PVdC/24 μ OPP/Acrylic	26	2.3	21
Amilon	Otto Nielson	15 μ OPA/60 μ LDPE	65	14.0	60
Melinex 813	ICI	12 μ PET	12	37	110
Bicor MB200	Mobil	24 μ OPP	24	500	1100

with food irradiated to a dose of 10 kGy (1 Gy, or gray = 1 J/kg) and also that there were no particular nutritional or microbiological problems at that level. Irradiation is achieved with electromagnetic radiation sufficiently high in energy to cause ionisation. Forms of ionising radiation used on food materials include: gamma-rays, which are given during the decay of radionuclides such as cobalt-60 or caesium-137; high-velocity electrons which are produced in electron beam machines by taking the electrons emitted from tungsten elements and accelerating them by applying high voltages; and X-rays, which have similar properties to gamma-rays and are given by directing high-velocity electrons into a heavy metal target. A description of the irradiation equipment and typical layouts of irradiation facilities has been given by Wilkinson *et al.* (59). The economical aspects of food irradiation have been reviewed by Deitch (60).

Food irradiation is now permitted in many countries but usually for limited groups of food items. The main purposes of food irradiation are inactivation of microorganisms, insect disinfestation, modification of the ripening or senescence of fruits and vegetables, inhibition of sprouting of certain vegetables, and modification of physical properties (59).

Farkas (61) has reviewed the microbiological safety aspects of irradiated foods, particularly covering the relationship between dose requirement and microbial load. Some data on nutritional quality of irradiated vegetables have been given by Wilkinson (62). Morris (63) has reviewed the post-harvest irradiation treatment of different vegetables (asparagus, garlic, mushrooms, onions, papayas and potatoes) in terms of the overall benefits relating to decay control, delayed ripening, sprouting inhibition and use of combination treatments. One important conclusion was that the use of irradiation alone is not a viable proposition to reduce the rate of microbial decay because if a sufficient dose to achieve a sterilisation is applied, the host fruit or vegetable is more sensitive to quality changes such as softening than the pathogens for which the treatment is intended (64). Irradiation will be more effective as part of a combination process which may also include a heat treatment and refrigeration. Buckley *et al.* (65) have concluded that in such a case the heat treatment is best applied after irradiation.

The field of combination treatments including an irradiation stage has not been extensively studied as yet, though some useful work has been carried out. Kramer (36) examined the combined effects of irradiation (0, 0.5, 1.0 kGy), sulphite washing (0, 1000 ppm Na_2SO_3), and storage temperature (4, 13 °C) on shelf-life of mushrooms and concluded that best results were given at the lower storage temperature and highest irradiation dose. Sulphite washing appeared to raise yeast, mould and bacterial counts after three days' storage. Sommer and Mitchell (67) and Moy (64) concluded, after studying a number of vegetables and fruits, that refrigeration was necessary to maximise the potential post-harvest life. Gnanayutham and Wilkinson (68) evaluated the effect of irradiation on typical salad vegetables (beansprouts, green peppers, radishes, carrots and cucumbers) which were then stored for different times at

2 °C. They concluded that the appearance and flavour of the vegetables remained largely unchanged after a dose of 1.8 kGy, but that texture, particularly in the case of cucumbers, was significantly affected. A lower dose of 0.5 kGy also caused textural changes, but within acceptable limits. Foster and Kilcast (69) indicated that irradiation of beetroot at 1.0 and 2.0 kGy followed by storage at 2 °C caused no reduction in quality. Howard and Buescher (70) examined the effects of gamma irradiation (0, 0.5, and 1.0 Gy) on firmness and on different compositional parameters of refrigerated cucumber pickles. They related softening of the mesocarp tissues exposed to 1.0 kGy, to altered solubility characteristics of pectic substances.

It can be concluded that irradiation as a preservation method for vegetables is not a panacea: the dose levels necessary to give a sterilisation effect also cause significant reductions in appearance and textural quality. The answer will, therefore, lie in combination treatments involving lower level doses of irradiation together with, for example, heat processing and chill temperature storage. The major hurdle to the application of food irradiation remains, however, consumer antipathy, despite the overwhelming body of scientific data which indicate that the process is safe and that irradiated food would reduce the incidence of food-borne, disease-causing bacteria and parasites (71).

6.4 Manufacturing methods

A manufacturing system for a chilled vegetable product would include some or most of the following unit processes, though not necessarily in the order given: washing; peeling; dicing/slicing/cutting; blanching; pH adjustment by dipping or blanching in acid; a_W adjustment by drying or mixing with humectant or both; dipping in preservative or surface coating solution; fermentation or addition of competitive microflora; heat processing; assembly; packaging with or without gas atmosphere; chilling; irradiation; chill storage.

As has already been mentioned, many of these unit processes would not be carried out due to legal or marketing associated reasons, or they simply would not be applicable to the particular product under consideration. Some of the more relevant unit processes will be discussed below.

6.4.1 *Preparation methods*

The pre-preparation and preparation equipment for chilled vegetables products are the same as those which would be used for frozen or sterilised products and are covered in other chapters of this book.

6.4.2 *Heat-processing methods*

Three techniques are available industrially for heat processing vegetables for chilled products, ex-pack pasteurisation, open-pack pasteurisation, and closed-pack pasteurisation (Figure 6.1).

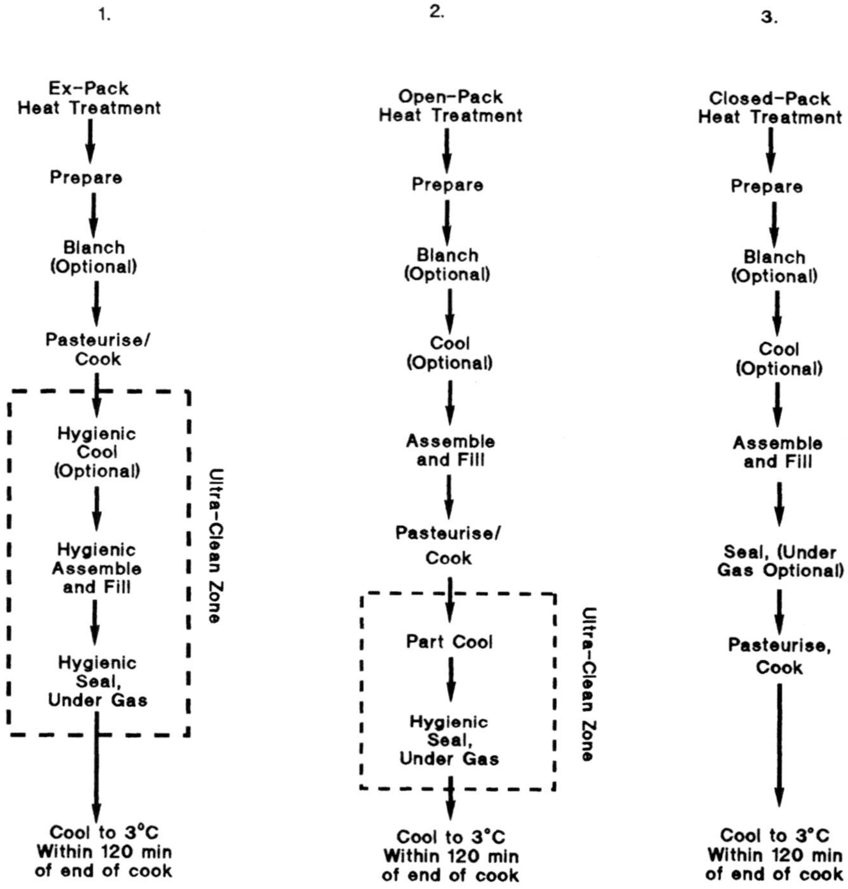

Figure 6.1 Methods for manufacture of heat-processed chilled vegetables: (1) Ex-pack heat treatment; (2) open-pack heat treatment; (3) closed-pack heat treatment.

The advantages of the first technique, pasteurisation away from the final package, is that vegetables can be separately cooked so that the optimum texture and organoleptic properties can be achieved in each. The individually cooked vegetables can then be cooled (optional) and merged, and introduced into the final package. The disadvantage of separate pasteurisation is that the risks from post-process contamination are greater than for the other two techniques, and steps need to be taken to reduce this risk. The usual method of reducing the risk is to assemble, fill and pack the product in a clean area – that is, an area in which steps have been taken to reduce the environmental microbial load in the air and on the operators (q.v.). The problem remains of the method of introduction of the pasteurised vegetable across the hygiene barrier and into the clean area. Some adaptation of standard available cooking or blanching plant is necessary to achieve this. With continuous equipment such as water blanchers, steam tunnels or thermoscrews, or microwave cookers, this problem is easily solved by arranging the exit from the unit to be on the clean side of the barrier. With push-through batch cookers equipped with separate entry and exit ports, the exit port can be arranged to be on the clean side of the hygiene barrier. Bewley *et al.* (72) describe a batch-type mobile closed cooker which can be disconnected from a heating and cooling service plant, and wheeled into a clean area before opening to the atmosphere.

With all of these systems, it is essential to monitor the time and temperature of the cook process, because for chilled foods the value of the process is usually a critical point in the assurance of safety (73). For continuous processes, careful consideration must be given to measure the value of the fastest moving particle. With microwave cookers, measurements are usually made using optical fibres tipped with heat-sensitive coloured beads (74).

Open-pack and closed-pack pasteurisation methods are applied to pre-assembled products. The former method allows greater flexibility in the choice of packaging format and simpler processing equipment. However, care has to be taken to protect the pack sealing surface from contamination by food debris during the cook process as this can cause problems later at the sealing point. Open-pack pasteurisation is achievable with any suitable belt or carrier pasteuriser, such as a steam tunnel or microwave tunnel. For obvious reasons, water cooking cannot be used.

Closed-pack pasteurisation is normally the system of choice, as it is inherently safer than the other two methods and does not require such extreme hygiene levels in the sealing area. Sophisticated batch pasteurisers are available which can provide close control of over-pressure, necessary particularly when the product is packed under MAP (75, 76). When utilising continuous, non-pressurisable, pasteurisers, the packaging material must be carefully chosen to be able to survive the process without undue evidence of distortion and without destroying the hermetic seal. Recently continuous microwave pasteurisers have been used more frequently, particularly those from AVP Magnetronics (UK) (77), Omac (Italy) (78) and Berstorff (Germany) (79).

6.4.3 *Chilling methods*

Basically, the loss of the heat from the food material is governed by the general principles of physics in that these losses occur due to four basic methods: radiation, conduction, convection and evaporative cooling. Practical methods of cooling food systems are based on convection and conduction. The rate of heat loss will depend on the temperature gradient, surface area and surface heat coefficient. The latter component is very important and is related to shape, size and topography, but not on the chemical composition of the food material. The velocity and the thermophysical properties of the heating medium are also important, for example, slow moving air, may have cooling values in the range of 5 W/m^2 °C as compared to 500 W/m^2 °C for agitated water. Another factor which may become critical is the heat transfer within the food particle, from the inside to the outside, and this is related to conductive properties.

There is less tendency to use cryogenic methods for chilled foods than for frozen foods because of the risks of surface freezing. Chilling systems using static or forced air, or using immersion or spraying with chilled water or brines are more often utilised (80).

Moving air is probably most commonly used because of economic and hygiene reasons; however, the process can cause dehydration of the surface and also freezing. Cold water misting is a method often used in supermarkets and also for cabbage and root crops. Hydrocooling is an efficient method when small pieces are involved. The vegetables are immersed or sprayed with cool water. The microbial quality of the water is important and disinfectants are often added to prevent spoilage. Such a system is used for celery, asparagus, corn, carrots and peas. Vacuum cooling is an extremely efficient process in those cases where the surface area is large and product can easily release internal water. The initial investment may be high. Generally, there is a 5 °C decrease in temperature for about 1% loss of moisture. Products may be sprayed prior to putting in the vacuum chamber (4 – 5 mmHg). The procedure is used for lettuce, celery, and cauliflower.

In all cases, as has been mentioned, the time and temperature conditions of the cooling process should be monitored.

6.4.4 *Sanitary packaging methods*

Some reference has already been made to the use of the fortress principle for hygienic manufacture and packaging of chilled food products. The assembly and sealing room will normally be the most critical zone and clean-room techniques are often employed therein. A description of clean-room principles and layouts has recently been given by Morel (81). A schematic diagram of a typical clean room applied to the assembly and packaging of chilled foods is given in Figure 6.2. The critical area over the filling line is serviced by filtered horizontal or vertical laminar air, typically of class 100 (that is, nor more than 100 particles of 0.5 μm size or higher in 1 cubic foot of air). The air emerges from the room through grills and ducts on the opposite side to the filling line, circulates back to a plenum chamber, is refiltered and enters the room again. Some provision is made for introduction of a small portion of fresh air to make up for losses through ports and doors. The air can be conditioned to a lower temperature than ambient (ideally 10–12 °C) and humidity control is practised.

Operators working in clean rooms are garbed with special overalls to reduce contamination of food or the room air by skin and other body particles.

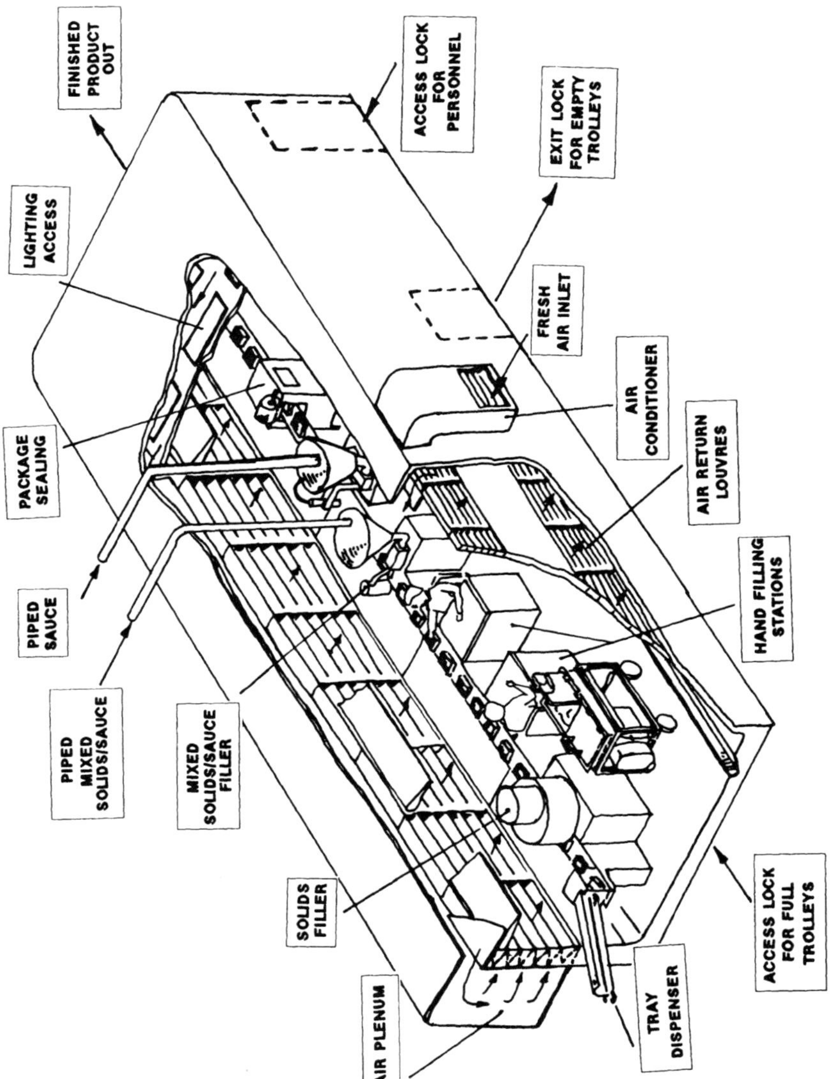

Figure 6.2 Clean room for filling and sealing chilled foods.

Operators wear mouth masks, hair and beard snoods, and gloves.

In the manufacture of chilled vegetable products, it will not be necessary in every case to utilise all the clean-room precautions and procedures to the degree described above. However, some elements will usually apply, including air overpressure with filtered air (not necessarily laminar), temperature controls, and hygienic garbing of operators.

6.4.5 *Manufacturing methods for catering outlets*

The unit processes described in the previous section are generally applicable to the manufacture of chilled vegetable products for retail or for food service outlets. In the field of food service, however, some rather more specific techniques have been developed over the years to satisfy the distinct requirements of the catering trade, such as multiportioning, convenience of regeneration and low labour costs.

Two systems in particular have gained wide use and these are the so-called cook-chill and *sous-vide* systems. Perhaps the most widely employed is the cook-chill system, used, for example, for production of hospital meals, school meals, in-flight meals and hotel meals. Over 300 in-house cook-chill installations have been identified in the UK alone (82). Guidelines on correct practices in cook-chill have been issued by the Department of Health and Social Security, UK (35). The microbiological and sensory quality of cook-chilled foods has been discussed by Robson and Collison (83) who concluded that, providing guidelines were followed, then meals were sound in terms of microbiological quality, but that, in certain cases, flavour and texture changes occurred within the recommended 5 days' storage.

The cook-chill system, as recommended by the DHSS, UK, is outlined in Table 6.2. Key points are cooking to a centre temperature above 70 °C under hygienic conditions, rapid chilling within a prescribed time, storage at 0–3 °C and regeneration to a temperature above 70 °C. Cook-chill thus employs ex-pack pasteurisation and hot filling. Cook-chill applies usually to multiportion high-volume preparations. Storage containers are either gastronomic trays

Table 6.2 Stages in the cook-chill process—the DHSS Guidelines.

Process stage	Time	Food temperature
Initial cooking	–	> 70 °C (centre temperature)
Portioning/packaging	< 30 min	–
Rapid chilling	90 min (or 2.5 h for meat joints up to 3 kg maximum weight)	0 °C to 3 °C
Refrigerated storage	5 days	0 °C to 3 °C
Tolerance	–	+ 2 °C for short periods
Distribution	–	3 °C (+ 2 °C as above)
Reheating	–	> 70 °C (centre temperature)

with tight-fitting lids, or barrier plastic casings which are closed by clipping. The advantages of cook-chill are that there is no handling of the raw food ingredients after they are added to the cooking kettle or oven, there are labour savings through the replacement of numerous small kitchens by one central facility and there is minimal product degradation due to the low processing temperatures. Zacharias (84) has reported that the main problem with cook-chill vegetable products is the development of off-colours after 2–3 days' chill storage.

Another technique employed for the manufacture of vegetable and other products for use in catering and institutional outlets is *sous-vide* processing (*sous-vide* is French for 'under vacuum'). Originally developed in France in the 1970s by the Chef George Pralus, this technique has recently gained popularity elsewhere in Europe and even in the USA. The process involves vacuum packing of prepared, raw or par-cooked meals or meal components, pasteurising them under strictly controlled time–temperature conditions (that is, closed-pack pasteurisation), chilling, then storage under refrigeration conditions (< 3 °C) until regeneration (85, 86). Similar advantages are claimed for *sous-vide* processed products as for cook-chill products: the minimal need for skilled labour at the regeneration facility (restaurant or canteen); the freshness characteristics resulting both from the minimal heat-processing conditions employed and from the capture of flavours and aromas normally lost during conventional heating; and the extended shelf-life, up to 40 days in some cases.

Interestingly, some of the principles of the *sous-vide* catering process system have more recently been applied to the manufacture of products such as ready meals based on meat or fish and vegetables, for the retail trade. This has occurred particularly in France, but also lately in the UK and USA. Products are vacuum packed in single- or double-serving trays or pouches, pasteurised, then cooled, usually in chilled water. It is not possible, on grounds of economy, to replicate the long-time low-temperature cook process often utilised for catering *sous-vide* products, and high-temperature short-time pasteurisation conditions are normal. In France, a licensing protocol for industrial and catering manufacture of vacuum-packed products is administered by the CNERPAC, a section of the Ministry of Agriculture. The protocol is based on different regulations (*arrêtes*), particularly those of June 1974 and September 1980 which describe hygiene requirements, cooking and chilling conditions, packaging conditions and storage temperatures (87, 88). Microbiological control of such products has been described by Poumeyrol *et al.* (89).

6.5 Distribution and retailing

It is important with all chilled foods, because of their short shelf-lives and perishable nature, that the supply chain is fast and efficient: transport and warehousing must, therefore, be at the correct temperatures and handovers

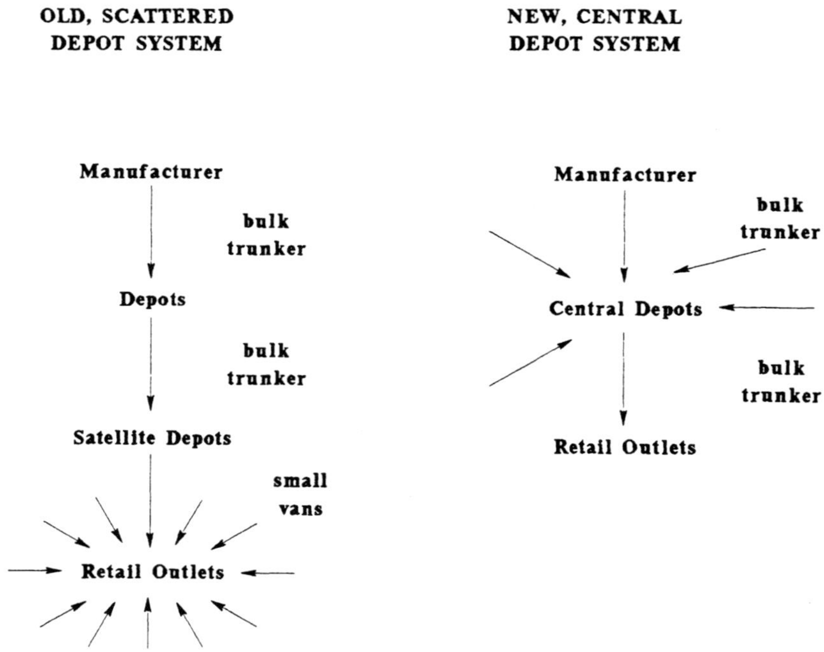

Figure 6.3 Chilled food distribution: the central depot system.

must be rapid and kept to a minimum. The Institute of Food Technology UK (90) has issued guidelines for the handling of chilled foods. Recommended temperature ranges relevant to vegetables are 0–5 °C for prepared salads and 0–8 °C for vegetables. Problems in the past with distribution chains have been identified as being associated with poorly organised handovers or with delivery vans on urban routes involving many drops (91). These types of problems relate more to the scattered depot style of distribution system which has now been largely replaced by central distribution in most countries in Europe and the USA. A comparison of the product flow in the two systems is shown in Figure 6.3.

In the retail phase, good temperature control, correct retail cabinet loading practices and short cabinet dwell times are again critical. A number of surveys in Europe and USA have shown that the temperatures at the point of sale of different chilled foods, including salad vegetables and prepared salads, can be very variable and sometimes well outside guidelines (92–94). This is despite recent improvement in retail cabinet technology. Temperature differences have been noted from the front to the back of cabinet shelves and from shelf-to-shelf (91, 92, 95). In food service applications involving vending machines supplying chilled salads and other meals, the picture is often no better. A recent survey in the UK (96) of eleven CFVMs of four types indicated considerable temperature fluctuation in different parts of many of the machines and an inability in some to maintain temperatures between the 0 and 5 °C recommended.

One other factor which has been reported to affect chilled food packs in retail is the so-called greenhouse effect. This is the entrapment within food packs of short wavelength radiation energy originating from the lighting systems in cabinets and stores. This energy then gradually heats the pack contents. It is probable that with modern packaging materials the effect is now uncommon (97).

Temperature abuse of chilled products in the chill chain is not clearly apparent until microbial growth has taken place through many log cycles to produce visual and olfactory effects. In some cases temperature abuse may not be apparent at all, even though the food may be in a risk state. One solution to this problem may be the use of time–temperature indicator tabs on each pack. The subject of time–temperature indicators, as they relate to chilled vegetable and other products, has been reviewed by Ballantyne (98), Campbell (99) and Wells and Singh (100). A time–temperature indicator is an inexpensive device capable of integrating the time–temperature history of the product by means of a mechanical, enzymatic, chemical or microbiological system which changes irreversibly from the time of activation. The change is usually represented as a visible response in the form of a colour development, a colour movement or a mechanical deformation. The visible response thus gives an indication of the storage conditions that have preceded it.

For many reasons the currently available integrator devices have not been taken up by food producers in significant numbers. The most important reasons are cost, reliability and concern for the implications in the market place. In connection with the latter, one problem which remains to be solved is the development of a consumer-friendly device which will indicate to the purchaser in a clear fashion how much time remains before the food will become unsafe or go below the optimum eating quality. There have recently been some improvements in the other factors mentioned, cost and reliability. Legislative demands may ultimately result in the more general application of time–temperature devices, to give a better indication of the state of the product than relying on the sell-by or consume-by date.

The final phase of the shelf-life of a chilled vegetable product is spent in the consumer's refrigerator prior to consumption. There is further opportunity for temperature abuse of such products during this phase. Surveys of consumer refrigerator temperatures have given results varying from good (94) to poor (101).

6.6 Nutritional and quality aspects

Quality losses in chilled vegetable products, as has been mentioned, usually occur as a result of chemical, enzymatic and microbial changes. The loss of food quality can be described in terms of the classical Arrhenius rate equation,

$$\text{Rate of loss} = k\ A^n$$

where A is the quality under study, k is a constant unique to each food component and n is the order of reaction. Enzymatic degradations are usually zero-order reactions (that is, $n = 0$), as are some lipid oxidations and non-enzymatic browning reactions. Microbial growth and vitamin losses are first-order, or higher, reactions.

Without some kind of heat treatment to inactivate enzymes, nutrient losses and quality changes in vegetables which have been minimally processed (washed, peeled, sliced) proceed rapidly. The consequences of mechanical damage, low-temperature storage and controlled atmosphere storage on the quality of raw fruits and vegetables have been thoroughly reviewed by Rolle and Chism (102). Other aspects such as the influence of cultivar type and stage of maturity at harvest have been reviewed by Shewfelt (5). Klein (103) has discussed and reviewed particularly the nutritional consequences of minimal processing.

Vitamin C is one of the most unstable vitamins with regard to losses during processing. It is, therefore, often used as an indicator of the nutritional value of vegetables. Fennema (104) examined vitamin C losses in several vegetables and indicated that losses increased when vegetables were sliced compared to whole. Normally, low-temperature storage retards metabolic changes, even so some decreases in nutrients are inevitable. Zepplin and Elvehjem (105), for example, showed 50% loss of ascorbic acid during six days' chilled storage of lettuce, while broccoli lost 25% under the same conditions. The effects of CA storage on vitamin and nutrient content are ambiguous and more work is required in this field to establish meaningful data.

Blanching and cooking of fresh vegetables also causes nutrient losses, particularly of vitamin C. Nevertheless, it is usually advantageous to carry out a blanch treatment in order to inactivate enzymes which cause deteriorative quality changes. The method of blanching is important and high-temperature short-time treatments give the best results. Nesheim (106) has indicated that vitamin C losses range from 0 to 12% after steam blanching, 5 to 36% after short-term water blanching and up to 94% after extended water blanching.

Blanching of vegetables also generally helps to retain textural and other sensory properties. Wei-Na (107), for example, demonstrated that blanched broccoli stored at 0 to 2 °C rated higher in sensory attributes than unblanched broccoli stored at the same temperatures for up to 30 days.

In summary, it is evident that chilled vegetable products will lose quality with time. Nutritional contents, flavour and aroma, appearance and texture will all decline due to enzymatic, microbial and chemical reactions. Because chilled foods are the closest that the industry can approach to kitchen-fresh prepared foods, they will always be compared to the home-made control. The determination of the essential sensory characteristics and hence the sensory shelf-lives of such foods, is, therefore, of great importance. This determination of acceptable sensory characteristics becomes more complex with multicomponent ready meal products which may contain meat or fish, sauce and vegetable constituents. Turner and Collison (108) recently reviewed some work in this area as part of an interesting study to examine the importance to consumers of different parts of the total meal, starter–entrée–potato–vegetable–dessert. Not unexpectedly, the entrée component was found to have the dominating effect in determining the overall meal acceptance, but the potato and vegetable components were also relevant.

Shelf-lives are normally determined by trained panellists and/or by consumer testing. Products are held for evaluation under conditions which will simulate the conditions through distribution, retail and the period under the consumer's care. In this respect good data on the expected time–temperature conditions throughout the chill chain must be available, as already discussed. The producer will then determine the limit of acceptability, taking also into account perhaps the economic aspects of the increased numbers of unsaleable product (that is, product beyond the sell-by date) associated with shorter shelf-lives.

6.7 Safety aspects

The storage life of chilled products, including vegetable products, is governed particularly by microbial safety considerations. As has been discussed, the microbial condition of chilled foods is dependent on the quality of the raw materials, the hygiene of the processing and handling operations, the correct control of certain process steps such as the heating and cooling stages, the presence of secondary barriers such as modified atmosphere, reduced pH or a_w, and finally on the maintenance of low temperatures during distribution, retail and storage in the consumer's refrigerator.

Attempts have been made to derive equations which model the growth characteristics of psychrotropic microorganisms at chill temperatures, in order that shelf-life predictions may be made. One of the most successful of these equations is that due to Ratkowsky *et al.* (109) in which a straight line is obtained when the reciprocal of the square root of the time required to produce a given increase of log microbial count is plotted against the absolute temperature. This equation has been shown to be applicable, for example, to chilled delicatessen salads (110), vegetable lasagne and vegetable pie (111). Baker and Genigeorgis (112) have reported the use of predictive modelling in quantifying the risk of toxicity in a large number of different types of *sous-vide* products, the data originally being based on that for raw fish.

A number of pathogenic bacteria have been shown to grow at refrigeration (0–4 °C) or slightly abused refrigeration (4–8 °C) temperatures. These include *Yersinia enterocolitica* (0 °C), *Listeria monocytogenes* (0 °C), *Aeromonas hydrophilia*, *Escherichia coli* (7–10 °C), various *Salmonella* sp. (6–7 °C), *Vibrio parahaemolyticus* (3.0 °C), *Staphylococcus aureus* (6.6 °C), non-proteolytic strains of *Clostridium botulinum* types B and F (3.3 °C), type E *C. botulinum*, (3.3 °C), *Bacillus cereus* (5–7 °C), *Clostridium perfringens* and *Campylobacter fetus*. The temperatures in parentheses refer to the lowest reported growth temperature for each microorganism.

Not all of these microorganisms are part of the characteristic flora of vegetables; however, if vegetables are only one of the components in a complex chilled meal product, it is possible at one time or another for almost all of the microorganisms listed to be present at some stage. General reviews on the microbial ecology of vegetables have been published by Deak (13) and by Hobbs (113). Brackett (11) has reviewed the microbiological consequences of minimally processed fruits and vegetables. Corlett (114, 115) has discussed microbiological safety considerations in refrigerated convenience foods.

Regarding specific food-poisoning organisms, Bialkowska *et al.* (111) have shown that *Y. enterocolitica* and *L. monocytogenes* could survive and grow at all temperatures tested (2–15 °C), if inoculated into a vegetable lasagne and a vegetable pie, but that *S. heidelburg*, *S. aureus* and *B. cereus* did not survive below 12 °C. Rose (116) showed that in delicatessen salads containing vegetables, proliferation of *C. perfringens, E. coli, Salmonella* sp. and *S. aureus* was largely prevented below 20 °C in high-acid products, below 10 °C in medium-acid products and below 4 °C in low-acid products. Lund and Brocklehurst (117) assessed the risks from food poisoning microorganisms in mayonnaise based salads and concluded that there is a potential risk from various microorganisms if the temperature is above 10 °C and the pH range is 4.8–5.0 or higher. Conner *et al.* (118) reviewed the literature relating to the risks from *C. botulinum* toxigenicity in chilled vegetable products and speculated that MAP cabbage might have been the source of toxin in one recent outbreak of botulism. Notermans *et al.* (119) reported that both proteolytic and non-proteolytic spores of *C. botulinum* could survive a typical process for potatoes of 95 °C for 40 min and then could grow and form toxin within six weeks at 10 °C. No toxin was detected at 4 °C. Cox (120) in a general perspective on *L. monocytogenes* noted that this microorganism might be a problem in longer shelf-life salads and vegetables.

The situation regarding the potential risks with chilled processed vegetable products and the precautions to be taken to minimise or avoid these risks can be summarised as follows. If the vegetables are prepared but uncooked or untreated in any other way, then the shelf-life is usually limited to a few days. The limitation is primarily due to physiological changes which cause losses in sensory quality. In raw products which are treated in a way that may tend to retard or disguise the physiological changes, for example, application of CAP, or reduction of pH in mayonnaise delicatessen salads, the shelf-lives usually become limited by the growth of certain non-pathogenic psychrotrophs such as different moulds or lactic acid bacteria. In some cases food poisoning psychotrophs such as *Pseudomonas* (121), *Salmonella* sp. or *Staph. aureus* have the potential to survive unless the temperature is very low or other conditions are unfavourable.

With cooked vegetable products, the cooking process is sufficient to remove the risks due to the non-sporeformers such as *Yersinia*, *Listeria*, *E. coli*, *Salmonella*, *V. parahaemolyticus* and *Staph. aureus*, because all of these are killed by even the mildest cook conditions. Nevertheless, care must be taken by the processor to ensure that an effective kill is given to all parts of the product and that there are no cold spots. Sporeforming pathogens, such as certain types of *C. botulinum*, will not be killed by mild cooking processes and other precautions need to be taken to minimise the potential for growth.

For products which are not cooked in the final package at point of manufacture, then there is the potential for recontamination by non-sporeforming psychrotrophs prior to packaging. Here the risks can be reduced by use of good manufacturing practices (GMP) and sanitary packaging procedures, and, for some pathogens, by MA packaging. Packing in atmospheres containing CO_2 has, for example, been shown to retard the growth of *Listeria* and *Yersinia* at chill temperatures compared to their growth in air (122).

Non-proteolytic B and F types of *C. botulinum* also have a very limited heat resistance and would be expected to be killed by an adequate cooking process. They do not survive at storage temperatures less than 3 °C. The proteolytic strains of B and F *C. botulinum*, and all strains of type A, will grow and produce toxin at 12 °C and above, under favourable conditions. These strains are highly heat resistant. The risks from these pathogens can be removed by proper temperature control of the storage phase, or by the utilisation of other hurdles (secondary barriers) such as reduction of the water activity or pH, and addition of antimicrobials.

In this connection, different workers have studied the interaction of a_w, pH and temperature on the growth of *C. botulinum*. Baird-Parker and Freame (123), for example, showed that at 30 °C and a pH of 7.0, the minimum a_w for growth was 0.95 for type A, 0.94 for type B and 0.97 for type E. At 20 °C, the limits were 0.97 for all three types. At 10 °C, the limit for type E was 0.99. Lund *et al.* (124) showed that *C. botulinum* type E grew and formed toxin at 8 °C and pH 7.0, but no growth occurred at temperatures between 8 and 20 °C at pHs below 5.1. Later work by Graham and Lund (125) indicated no growth of *C. botulinum* at pH $\geqslant$ 5.4 and 16 °C or at pH $\geqslant$ 5.5 and 12 °C. Another method of reducing the risks from *C. botulinum* hazards is by biocontrol, that is, the incorporation of an acid-producing microorganism in the food product which will grow under abuse conditions to give sufficient acid to prevent toxigenesis by the pathogen. Tanaka *et al.* (126) utilised lactic acid bacteria for this purpose.

For chilled food products which are reheated by the consumer prior to eating, the kitchen reheating stage does not necessarily completely remove the risks from any non-sporeforming pathogens which may have survived the manufacturing process or have recontaminated the product. In recent work carried out by Walker *et al.* (127) it was demonstrated that when a number of ready meal products, some containing vegetables, were challenged with large populations of *L. monocytogenes*, these microorganisms could, in certain cases, survive the recommended microwave oven reheating conditions. The reason for this was the uneven heating characteristics of the products in the microwave oven, which meant that in particular locations in the samples lethal temperatures for *L. monocytogenes* were simply not attained.

The conclusion from the above discussion is that the production of refrigerated vegetable products with extended shelf-lives should take place in well-controlled and hygienic plants, and should be based on scientific knowledge of all the interactive factors which can assure safety of the products at the point of consumption. The primary factor in this situation is, of course, the maintenance of an adequately low temperature throughout the chill chain.

6.8 Conclusions

Consumer trends towards prepared value-added convenience foods, associated with other demographic and market trends, should continue to favour the development and increased sales of chilled vegetable products. Raw prepared and cooked ready-to-eat or ready-to-heat products can be manufactured by modern processing methods, but high hygiene levels, good GMP and correct HACCP (hazard analysis critical control point) procedures must be in place to ensure that products are of the optimum quality when they leave the chilled food factory. Maintenance of the chill chain at low temperatures must then be practised throughout the shelf-life of the product. Despite this, there is the potential for vegetable product deterioration due to microbial growth, nutrient losses and general quality changes. Some of the methods available to prevent or retard these deteriorative effects have been described, such as reducing temperatures to around 0 °C, gas packing or manipulating product recipes to reduce a_w or pH. Some rather controversial methods such as *sous-vide* processing or irradiation treatment may gain more favour with the consumer and the legislature in the future.

References

1. Young, J.N., in FAST Report: *The Future of the Food System (Alim 1): The Impact of Food Science and Technology on the EEC Food Industry in the Next 10–15 Years*, Leatherhead Food Research Association, 1987.
2. Shewfelt, R.L., *J. Food Qual.* **10** (1987), 143.
3. Rolfe, E.J., 'The chilling and freezing of foodstuffs', in *Biochemical and Biological Engineering Science*, Vol. 2 (ed. N. Blakebrough), Academic Press, New York, 1968, p. 138.
4. Huxsoll, C.C. and Bolin, H.R., *Food Technol.* **43** (1989), 124.
5. Shewfelt, R.L., *Food Technol.* **40** (1986), 70, 89.
6. Leistner, L. and Rodel, W., in *Inhibition and Inactivation of Vegetable Microbes* (eds F.A. Skinner and W.B. Hugo), Academic Press, London, 1976, p. 219.

7. Plank, R., *Kaltetechnik* **11** (1959), 306.
8. *Sanyo Tech. Review (Japan)*, **20** (1) (1988), 116.
9. Wang, C.Y., *Food Rev. Int.* **5** (2) (1989), 209.
10. Jackman, R.L., Yada, R.Y., Marangoni, A., Parkin, K.L. and Stanley, D.W., *J. Food Qual.* **11** (1988), 253.
11. Brackett, R.E., *J. Food Qual.* **10** (1987), 195.
12. Chirife, J. and Ferro Fontan, C., *J. Food Sci.* **47** (2) (1982), 661–663.
13. Deak, T., 'Microbial-ecological principles in controlled atmosphere storage of fruits and vegetables', in *Microbial Associations and Interactions in Foods* (eds I. Kiss, T. Deak and K. Incze), Reidel, Dordrecht, 1984, pp. 9–22.
14. Kester, J.J. and Fennema, O.R., *Food Technol.* **40** (12) (1986), 47.
15. Hardenburg, R.E., *Wax and Related Coatings for Horticultural Products: A Bibliography*, Agricultural Research Service Bulletin, 51–15, USDA, Washington, D.C., 1967.
16. Hartman, J. and Isenberg, F.M., *Waxing Vegetables*, New York Agricultural Extension Service Bulletin No. 965, Cornell Univ., Ithaca, New York, 1956.
17. Unilever Ltd, UK Patent 2,016,042A, 1979.
18. Lowings, P.H. and Cutts, D.F., *Proc. Inst. Food Science Tech. Ann. Symp., Nottingham, England*, July 1981, p. 52.
19. Everest-Todd, S., United States Patent 4,532,156, 1985.
20. Kirsop, B.H. and Brocklehurst, T.F., European Patent Application, EP 9 253 535 Al, 1988.
21. Baardsath, P., *NINF Informasjon*, No. 1 (1980), 42.
22. Rice, J., *Food Process. USA*, **44** (12) (1983), 74.
23. Miller, W.R., Spaling, D.H. and Risse, L.A., *Proc. Florida Horticultural Society* **96** (1983) 347.
24. Kratky, Z. and Vadehra, D.V., US Patent 4,810,512, March 1989.
25. Goussault, B., French Patent 2,589,330, 1987.
26. Passam, H.C., *Int. Agric. Devel.* **2** (9) (1982), 10.
27. Malik, M.N., Khan, M.A. and Siddique, M., *Scientia Horticulturae* **10** (4) (1979), 331–336.
28. Drake, S.R. and Lambert, E.F., *J. Food Qual.* **8** (2/3) (1985), 101.
29. Aderiye, B.I., *Food Chem.* **16** (2) (1985), 69.
30. Tomita Seiyaku, K.K., Japanese Patent 5,530,341, 1980.
31. Olarunda, A.O. and Kitson, J.A., *Food Product Devel.* **11** (4) (1977), 44, 46.
32. Batal, K.M., Heaton, E.K., Granberry, D.M. and Benchat L.R., *J. Food Sci.* **47** (5) (1982), 1675.
33. Shapton, D.A., Lovelock, D.W. and Laurita-Longo, R., *J. Appl. Bact.* **34** (2) (1971), 491.
34. Catteau, M. and Poncelet, F., *Proc II Int. Congress of Food Science and Technology*, Vol. III (1974), p. 114.
35. Department of Health and Social Security (UK), *Guidelines on Pre-Cooked Chilled Meals*, HMSO, London, 1980.
36. Stevens, R.L., *Symp. Proc. Campden Food and Drink Research Association*, 7–9 May 1985, pp. 3–8.
37. Livingston, R.E., *J. Foodservice Systems* **3** (1985), 221.
38. Kidd, F. and West, C., *Report of the Food Investigation Board*, London, 1936.
39. Kidd, F. and West, C., *Report of the Food Investigation Board*, London, 1932, p. 209.
40. Jurin, V. and Karel, M., *J. Food Sci.* **17** (1963), 104.
41. Labuza, T.P. and Karel, M., *20.52 Laboratory Procedures*, MIT, Department of Food Science and Nutrition, 1967.
42. Tompkins, R.G., *Food Manuf.* **42** (4) (1967) 34.
43. Rizvi, S., *Crit. Rev. Food Sci. Nutrit.* **14** (1981), 11.
44. Ben-Yehoshua, S., *Proc. Int. Soc. Citriculture* (*Australia*), 1983, p. 110.
45. Labuza, T.P. and Breene, W.M., *J. Food Process. Preserv.* **13** (1989), 1.
46. Barmore, C.R., *J. Food Qual.* **10** (1987), 207.
47. Day, B.P.F., Campden Food and Drink Research Association, Technical Memorandum No. 524, 1989.
48. Day, B.P.F., 'A perspective of modified atmosphere packaging of fresh produce in Western Europe', in *Proc. of CAP 90*, Schotland Business Research Inc., 1990, pp. 35–54.
49. DeLassus, P.T. and Hilker, B.L., in *Food Product-Package Compatibility Proc.* (ed. J.L. Gray, B.R. Harte, and J. Miltz), Tecnomic Publishing Co. Inc., 1986, p. 229.

50. Labuza, T.P., in *Shelf Life Dating of Foods*, Food and Nutrition Press, Conn., 1982.
51. Simon, I., Labuza, T.P. and Karel, M., *J. Food Sci.* **36** (1971), 280.
52. Ballantyne, A., Stark, R. and Selman, J.D., *Int. J. Food Sci. Technol.* **23** (4) (1988), 353.
53. Ballantyne, A., Stark, R. and Selman, J.D., *Int. J. Food Sci. Technol.* **23** (3) (1988), 262.
54. McLachlan, A. and Stark, R., Campden Food and Drink Research Association, Technical Memorandum No. 412, 1985.
55. Geeson, J.D., Browne, K.M., Maddison, K., Shepherd, J. and Guaraldi, F., *J. Food Technol.* **20** (1985), 339.
56. Mitsubishi, United States Patent 4,421,235, 1983.
57. Carnation, United States Patent 2,825,651, 1964.
58. The Report of a Joint FAO/IAEA/WHO Expert Committee, *The Wholesomeness of Irradiated Food*, Technical Report Series 659, WHO, Geneva, 1981.
59. Wilkinson, V.M., Hook, A.R. and Ley, F.J., British Food Manufacturers Ind. Res. Association, Technical Notes, No. 40, June 1986.
60. Deitch, J., *Crit. Rev. Food Sci. Nutrit.* **17** (4) (1982), 307.
61. Farkas, J., *Int. J. Food Microbiol.* **9** (1) (1989), 1–15.
62. Wilkinson, V.M., British Food Manufacturing Ind. Res. Ass., Scientific and Technical Surveys, No. 151, 1985.
63. Morris, S.C., *Food Technology in Australia* **39** (7) (1987), 336.
64. Moy, J.H., 'Radurization and radiation: fruits and vegetables', in *Preservation of Food by Ionizing Radiation*, Vol. 3 (eds E.S. Josephson, and M.S. Peterson), CRC Press, Boca Raton, 1983, pp. 83–108.
65. Buckley, P.M., Sommer, N.F., Coon, D.A., Dolby, M. and Maxie, E.C., *Radiation Bot.* **40** (1969), 26.
66. Kramer, M.E., Doores, S. and Beelman, R.B., The Effect of Radiation Processing on Mushroom Shelf-Life at Two Storage Temperatures, in *Changing Food Technology* (eds. M. Korger and R. Shapiro), Technomic Publishing Co., 1987, p. 165.
67. Sommer, N.F. and Mitchell, F.G., *Hortic. Sci.* **21** (1986), 356.
68. Gnanayutham, P.J. and Wilkinson, V.M., British Food Manufcturing Ind. Res. Assoc., Research Reports, No. 523, October 1985.
69. Foster, T.E. and Kilcast, D., British Food Manufacturing Ind. Res. Assoc., Research Reports, No. 616, March 1988.
70. Howard, L.R. and Buescher, R.W., *J. Food Sci.* **54** (5) (1989), 1266.
71. American Council on Science and Health, *Food Irradiation: A Brief Overview* (3rd edn), 1995 Broadway, New York, December 1988.
72. Bewley, D., Brimelow, C.J.B. and Wright, K., UK Patent Pending, Application No. 8913738.4, 1989.
73. Campden Food and Drink Research Association: HACCP Working Group, *Guidelines in the Establishment of Hazard Analysis Critical Control Points HACCP*, Technical Manual No. 19, 1987.
74. Richardson, P., Campden Food and Drink Research Association, Technical Bulletin No. 58, August 1985.
75. Peyremorte, J.-P., *IAA* (April 1988), 301.
76. Pinot, J.-P., *IAA* (April 1988), 305.
77. Burfoot, D., Griffin, W.J. and James, S.J., *J. Food Engng* **8** (1988), 145.
78. Frampton, R., *Food Manuf.* (May 1988), 53.
79. Byrne, M., *Food Manuf.* (May 1988), 57.
80. Campbell-Platt, G., *Food Technol. Int. Europe.* (1987), p. 63.
81. Morel, F., *IAA* (April 1988), pp. 289–293.
82. Walker, A.E. and Light, N.D., *Hospitality* **86** (1988), 15.
83. Robson, C.P. and Collison, R., *Catering and Health* **1** (1989), 151.
84. Zacharias, R., 'Chilled meals: sensory quality' in *Advances in Catering Technology* (ed. G. Glew), Applied Science Publishers, London, 1980, pp. 409–417.
85. Schafheitle, J.M. and Light, N.D., *Int. J. Food Sci. Technol.* **24** (1989), 199.
86. Dorset Institute of Higher Education, UK *An Introductory Handbook to the Cryovac* Cuisine en Papillote Sous-Vide System*, 1986.
87. Rosset, R. and Lameloise, P., *IAA* (April 1988), 241.
88. Beaufort, A. and Giuliani, L., *IAA* (April 1988), 245.

89. Poumeyrol, G., Rosset, R., Giuliani, L., Rosset, P. and Rudelle, S., *Sci. Aliments* **7**, Series VII (1987), 209.
90. Institute of Food Science and Technology (UK), *Guidelines for the Handling of Chill Foods*, IFST (London), 1982.
91. Brimelow, C.J.B., 'Temperature and time-temperature surveys of chilled foods in the UK', in *Proc. COST91-bis: Sub-Group Chilling Meeting* (ed. T. Gormley), Clermont Ferrand, France, May 1987.
92. Rose, S., Campden Food and Drink Research Association, Technical Memorandum No. 423, 1986.
93. Anon., *50 Millions de Consommateurs*, No. 201 (December 1987), 68.
94. Harris, R.D., *Food Process.* (December 1989), 111.
95. Bogh-Sorensoen, L., 'Temperatures in open display cabinets', in *Proc. 17th European Meeting of Meat Research Workers*, Meat Research Institute, Bristol, UK, 1971, pp. 491–495.
96. Light, N., Young, H. and Youngs, A., *Food Sci. Technol. Today* **1** (4) (1987), 252–256.
97. Gill, J., *Food Flav. Ingred. Pack. Process.* **10** (4) (1988), 47–51.
98. Ballantyne, A., Campden FDRA, UK, Technical Memorandum No. 473, January 1988.
99. Campbell, L.A., 'Use of temperature indicators in monitoring quality of refrigerated salads', M.S. Thesis, Michigan State Univ., 1986.
100. Wells, J.H. and Singh, R.P., *J. Food Sci.* **53** (1) (1988), 148.
101. Torrey, G.S. and Marth, E.H., *J. Food Protec.* **40** (6) (1977), 393.
102. Rolle, R.S. and Chism, G.W., *J. Food Qual.* **10** (1987), 157.
103. Klein, B.P., *J. Food Qual.* **10** (1987), 179–193.
104. Fennema, O., *Food Technol.* **31** (12) (1977), 32.
105. Zepplin, M. and Elvehjem, C.A., *Food Res.* **9** (1944), 100.
106. Nesheim, R.O., *Federation Proc.* **33** (11) (1974), 2267.
107. Wei-Na, L., 'Sensory quality and nutrient stability of raw and processed foods in storage', Ph.D. Thesis, Univ. of Maryland, 1981.
108. Turner, M.T. and Collison, R., *Food Qual. Pref.* **1** (1) (1988), 21.
109. Ratkowsky, D.A., Olley, J., McMeekin, T.A. and Ball, A., *J. Bact.* **149** (1982), 1.
110. Stannard, C.J. and Williams, A.P., Leatherhead Food Research Association, Res. Rep. No. 552, 1986.
111. Bialkowska, A., Noel, A., Williams, A.P. and Gibbs, P.A., Leatherhead Food Research Association, Res. Rep. No. 632, 1988.
112. Baker, D.A. and Genigeorgis, C., 'Microbiological risk assessment of refrigerated foods: the safety of 'sous-vide' products, *5th Int. Symp. World Association of Veterinary Food Hygienists, Stockholm, Sweden*, 2–7 July 1989.
113. Hobbs, G., *Microb. Ecology* **12** (1986), 15.
114. Corlett, D.A., *Cereal Foods World* **34** (12) (1989), 980.
115. Corlett, D.A., *Food Technol.* **43** (2) (1989), 91.
116. Rose, S., Campden Food and Drink Research Association, Tech. Memorandum No. 417, 1986.
117. Lund, B. and Brocklehurst, T.F., *Symp. Proc. Campden Food and Drink Research Association*, 7–9 May 1985, 145–152.
118. Conner, D.E., Scott, V.N., Bernard, D.T. and Kautter, D.A., *J. Food Safety* **10** (1989), 131.
119. Notermans, S., Dufrenne, J. and Keybets, M.J.H., *J. Food Protec.* **44** (1981), 572.
120. Cox, L.J., *Food Technol.* **43** (12) (1989), 52.
121. Kraft, A.A. and Rey, C.R., *Food Technol.* **33** (1) (1979), 66.
122. Genigeorgis, C., Principles of Preservation by Refrigeration, Modified Atmosphere and Moisture Control, Proceedings of the 1989 IFT Short Course, 'Minimally Processed Refrigerated Foods', 13–39, 1989.
123. Baird-Parker, A.C. and Freame, B., *J. Appl. Bacteriol.* **30** (1967), 420.
124. Lund, B.M., Wyatt, G.M. and Graham, A.F., *Food Microbiol.* **2** (1985), 135.
125. Graham, A.F. and Lund, B.M., *J. Appl. Bacteriol.* **63** (1987), 387.
126. Tanaka, N., Traisman, E., Lee, M.H., Cassens, R.G. and Foster, E.M., *J. Food Protec.* **43** (1980), 450.
127. Walker, S.J., Bows, J., Richardson, P. and Banks, J.G., Campden Food and Drink Research Association, *Technical Memorandum No. 548*, 1989.

7 Other preservation methods

KEITH ANDERSON

7.1 Introduction

This chapter covers those methods of processing vegetables for future consumption or use as food recipe ingredients which are not dealt with elsewhere in this book. There will, no doubt, still be some omissions, but the objective has been to include such processes as drying, salting or brining and halophilic fermentation, pickling, other water activity-related processing, and combination or 'hurdle' techniques. For further background reading, the publications of Luh and Woodroof (1), Duckworth (2) and Desrosier (3) are commended.

7.2 Dehydration

Drying is one of the longest established methods of food preservation, and one which occurs naturally, for example, with seeds. It combines the benefits of microbiological and physiochemical stability with reduction in weight and transport costs and has other advantages in handling and storage. It is of wide application, covering fish and meats, fruits and vegetables, and many part or wholly prepared food products. The list is endless, but examples of dehydrated foods include tea, coffee, granulated sugar, soft fruits, vine fruits, starches, legumes, root vegetables, soups, ready-meals, and herbs and spices. In this chapter the drying of extracts and infusions, such as coffee, and of juices has not been included.

The success of any drying operation depends on removing enough moisture from the food to achieve a water activity too low to allow microbiological growth to take place. This in turn means that there must be sufficient input and transfer of heat to provide the latent heat of vaporisation needed, and that the water or water vapour moves through the food and then away from it to separate the water from the foodstuff. As all moisture must be lost from exposed surfaces, normally to the air, it follows that particle size and geometry, and the relative humidity of the air used for drying are critical factors, as will be the manner in which the food is supported during the drying operation, that is, the extent to which potential exposed surfaces are covered by contact with other food particles, trays, or conveyors.

The complex physiological construction of most vegetables (and foodstuffs in general) restricts the rate of migration of water from within to the exposed outer surfaces from whence it can evaporate. Further, as evaporation takes place, tissue fluid in cells local to the surface becomes more concentrated, creating a moisture gradient across the material. As diffusion becomes progressively more difficult, structural failure occurs both in the individual cells and at a macroscopic level as evidenced by cracking, shrinking and shrivelling. Excessively fast removal of moisture from outer layers can result in case hardening effectively sealing the surface and causing inadequate drying in the centre.

Factors which affect the efficacy and the efficiency of drying operations are complex, and the reader is referred to the appropriate literature for more information on the engineering and physics involved. Earle's *Unit Operations in Food Processing* (4) provides a very good introduction to these aspects. Dalgleish (5) suggests that to determine optimum method and speed of drying the following should be known.

(i) The characteristics of air/vapour mixtures at different temperatures.
(ii) The characteristics of the material being dried:
 - its water activity at different moisture contents for a given temperature
 - its heat conductivity
 - its diffusion resistance
 - its effective pore size
 - any marked significant change points in the drying rate.
(iii) The rehydration quality or reconstitution characteristics expected after a defined period of storage.

Clearly, any method of agitating or turning the material during drying aids efficiency and uniformity of drying. In vegetable drying, product temperatures should not exceed 65 °C if adverse reactions are to be avoided.

7.2.1 *Water activity*

The ratio of the vapour pressure above a solution (P) (or complex mixture) to that above pure water (P_0) at a constant temperature is known as the water activity (a_w)

$$a_w = P/P_0.$$

This figure is most closely related to actual water availability for microbiological growth. Thus growth, and hence spoilage, can be predicted. Most bacteria do not grow below a_w 0.91, yeasts below 0.88, and moulds below 0.80. Exceptionally tolerant specialist organisms such as halophilic bacteria and osmophilic yeasts can grow down to about 0.60. Reference to the characteristics of potential spoilage organisms, therefore, will indicate the maximum a_w and hence moisture content that must be achieved in the drying process.

7.2.2 *Field drying and sun drying*

The drying of crops pre-harvest or while they are still in the immediate proximity of the growing area, for example, grain, hay, peas, is the historic root of preservation by drying, and situations where the combined effects of sun and wind provide the motive forces for effective evaporation may be referred to as field drying. Effective successful field drying depends on local climatic conditions at point of harvest and, therefore, is restricted to warm/hot parts of the world where wind and air humidity are satisfactory. Sun drying of harvested materials is variable and only possible where the climate is suitable. Fruits and seeds are more frequently dried than vegetables, but some fruits regarded as vegetables are dried successfully in this way, for example, peppers and the Sudanese sliced tomato salsa.

7.2.3 *Air drying*

In most drying situations mechanical assistance is needed over and above the forces of nature, this being the conventional concept of air drying. For effective drying the partial pressure of water vapour in the air surrounding the material to be dried must be significantly lower than its saturated partial pressure at the operating temperature. This can be expressed as a relative humidity ratio, and the lower the ratio the more effective the drying. As the ratio rises so fresh lower humidity air must be introduced into the system, or the water vapour removed from it by some means. The accurate measurement of equilibrium relative humidity (erh) is clearly essential as is that of temperature, and suitable methods are referred to in the literature (24).

Air drying may be batchwise or continuous and equipment includes tunnels (Figure 7.1), trays or oven dryers, drum or roller dryers, pneumatic, trough, rotary, cascade, tower, spiral, fluidised bed, bin, and belt or band dryers (Figure 7.2). Spray dryers, freeze dryers and vacuum dryers involve rather more than just air-drying principles, and are considered separately below. While the differences in these dryers cannot be elaborated in this chapter, most modes of operation are obvious by their names. Drying equipment for food has been well reviewed by van Arsdel *et al.* (6). In general hot-air dryers are designed to allow a high airflow in the early stages but not so much as to move the material being dried, except where this is a requisite feature as for fluidised bed dryers. Typically, an airflow rate of 180 to 300 m/min is used for vegetable pieces, with dry-bulb air temperatures of 90–100 °C and wet bulb temperatures below 50 °C. As moisture content falls the airflow rate is reduced and drying temperature is decreased to 55 °C or less until the moisture content

Figure 7.1 A tunnel type tray dryer (courtesy APV Pasilac Ltd).

is below 6.0%. Separate driers may be used to complete the final stages of drying.

In fluidised bed and airlift or pneumatic driers, the airflow must be of sufficient velocity to lift the particles to be dried from a porous plate bed and cause them to behave as a liquid. This allows them to flow over a weir to exit the drier. Movement within the airstream provides excellent mixing and hence even drying. This can be used for powders, dried vegetables and for small whole vegetables such as peas. Again, the moisture content and temperature of the air can be varied, as can the airflow to suit the material being dried.

7.2.4 *Spray drying*

In spray-drying systems a spray of the material to be dried, which might, for example, be a puréed vegetable suitably diluted and stabilised, is dispersed as an atomised spray into a countercurrent of hot dry air. The droplets dry in the airstream, falling to the bottom of the unit as they do so. Key elements are sufficient fan power and heating to achieve the required velocity and temperature, the atomiser, the drying chamber and the means of removing the

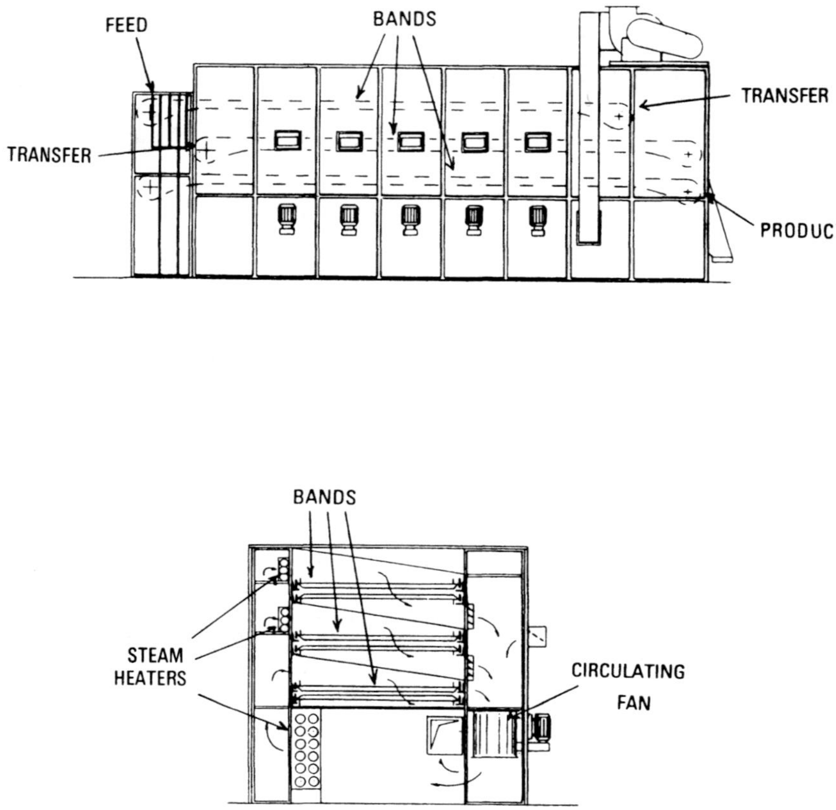

Figure 7.2 Cross sections of a three-pass band dryer (courtesy APV Pasilac Ltd).

dried product (3, 7). The atomiser may be one of three basic types: the simple jet, the dual outlet nozzle which uses air or steam together with the fluid to atomise the fluid, and the spinning disc on to which the feedstock is fed.

The main advantage in spray drying is the speed of drying, while a major disadvantage may be the large size of drying chamber.

7.2.5 *Vacuum drying*

Vacuum driers may be cabinet batch driers or continuous band driers with vacuum locks at infeed and outlet. In either case heat transfer is by radiation and conduction. The advantage of vacuum drying is that evaporation of water takes place more readily at lower pressures.

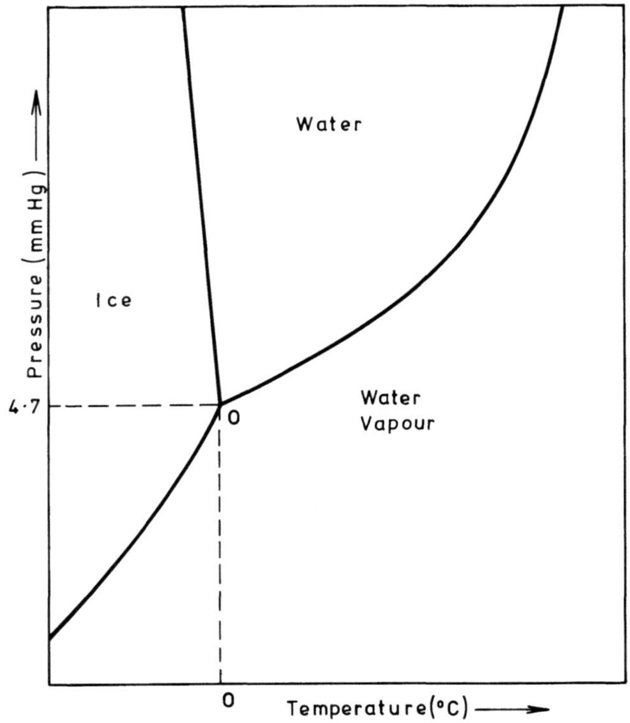

Figure 7.3 Water phase diagram.

7.2.6 *Freeze drying*

Freeze drying is a phenomenon which occurs when water is removed by direct vaporisation from ice. Water exists in three states and reference to the triple-point diagram for water (Figure 7.3) shows that within the right conditions of temperature and pressure this is what occurs. The triple point (0) is the point at which all three physical states can exist together, and for water is 0.0098 °C. By keeping the temperature and pressure below the triple-point conditions, direct vaporisation from ice occurs. This process minimises physical damage to the vegetable, enhances reconstitution characteristics and minimises the occurrence of oxidation and thermal processing reactions.

The raw material may be presented in the frozen state or freezing achieved by the evaporative process occurring at a low enough pressure. Temperature and pressure must be controlled to prevent the system allowing water to enter the liquid phase. It is, in fact, a development of vacuum drying and it may be effected in batch or continuous systems, heat input being by radiation and conduction from heat transfer plattons within the drier. Accelerated freeze drying (AFD) refers to the method of speeding the drying by positioning the material being dried between heated plates.

7.2.7 *Dehydrofreezing*

This is a composite process in which dehydration removes about half the water originally present, and the resulting material is quick frozen. Belt, trough and pneumatic dryers have been used and uniform drying of each and every piece of material is essential. This hybrid process has the advantages of greatly reducing the time required for the drying and rehydration processes and reducing, by about half, the storage space required for the frozen product, but the product has a shrivelled appearance which may require explanation to the consumer.

7.2.8 *Intermediate moisture food and osmotic drying*

Intermediate moisture foods are those which do not have a low enough a_w to ensure preservation by this means alone but are stable after combination with another means of preservation, for example, dehydrofreezing. Drying of some sort is usually needed to achieve this, but vegetables are not significant in this area other than after secondary or tertiary processing, for example, extruded soya products.

Osmotic drying has been practised, primarily with fruit, and involves the removal of water by osmosis through treatment of the feedstock with successively increasing concentrations of sugar solutions.

7.3 Dried vegetables

It is important that the raw material used is suitable for dehydration and will give the required quality on rehydration. This means the choice of suitable cultivars and harvesting at the right stage of maturity. Choice of cultivar may be influenced by sensory characteristics, suitability for mechanical harvesting and subsequent handling, suitability for drying, and rehydration performance as well as the normal considerations of yield, growing area, infection resistance, hardiness, and sowing and harvesting dates.

In general, vegetables for drying may be considered in four groups; legumes, root vegetables, brassicas and others. In broad terms the initial preparation is virtually identical to the preparation for canning or freezing and so these operations are not described in depth here. Preparation requires the removal of extraneous matter, rejection of defective vegetables, sorting and grading where appropriate, and possibly snibbing, peeling, slicing, dicing and other processes.

7.3.1 *Preparation*

Earth, stones and other detritus not removed in the field from root crops (such as beetroot, carrots, potatoes and onions) are dealt with by stone traps, horizontal rod or drum washers, or other similar devices. Grading may be achieved by a variety of methods such as diverging belts or vibrating mesh sieves, and this can be an important factor in ensuring uniformity of processing treatment in the later stages of peeling or cooking operations.

Apart from the obvious possibility of hand peeling, modern industrial peeling systems include steam, abrasion, lye, and flame with mechanical knife peeling for onions. Each system has its own merits and disadvantages, particularly in relation to yield, effluent production and amount of hand finishing required. In every case quality and yield are optimised where uniformly graded and regularly shaped feedstock is presented to the system.

Cutting, slicing, dicing and shredding operations may be carried out to give a wide variety of finished product and the chapter on preparation procedures should be consulted on these aspects.

Blanching is an essential pre-dehydration process for many vegetables to inactivate enzymes which would otherwise cause deterioration of colour, flavour and texture during drying and subsequent storage. Blanching is simply an elevation of temperature to achieve the required degree of enzyme inactivation and may be carried out by hot water scalding, steam or microwave heating. It involves heating to temperatures of 90 °C or more for relatively short periods of time (from 20 s to 2 or 3 min). During water blanching or subsequent water cooling additional treatments may be imposed such as the addition of sodium metabisulphate and sodium carbonate to give slight alkalinity to green vegetables to prevent phaeophytin formation. With water blanching the medium may also contain salt and pH regulators. Different types of blanching equipment are available for both batch and continuous treatments. Scalding may achieve the objectives of a blanching operation but is intended primarily to achieve a partial cooking of the vegetables.

7.3.2 *Legumes*

Apart from direct marketing for retail and catering, dried legumes have a significant role as ingredients in both dry and reconstituted forms, not only in dried food mixes such as soups and ready meals, but also for canning and for canned and frozen composite products. Examples include processed peas, butter beans and lentils.

Crop maturity at harvest is a critical factor and is assessed by different methods of which the Tenderometer has probably been the most widely used. Green beans are partially cooked by blanching prior to drying in appropriate forms, and chemical treatment such as sulphiting may also be applied prior to drying. Peas are blanched and the skins may be punctured before drying to assist removal of water and eventual rehydration.

In continuous hot-air drying a final target moisture content of approximately 5.0% is sought which is low enough to inhibit Maillard browning reactions. Puffing techniques may be applied towards the end of the drying process whereby the partially dried material is rapidly superheated and the pressure removed suddenly allowing the released steam to disrupt the cell structure and create passages speeding both the last stages of drying and eventual reconstitution.

For freeze drying the material must be frozen after blanching and this may be achieved by conventional means, or by evaporative cooling within the drier, possibly with the product in contact with refrigerated shelves. The freeze-dried material will have less crush resistance than its air-dried counterpart and the packaging should reflect this fact. Inert gas flushing is frequently employed when packing freeze-dried products.

7.3.3 *Brassicas*

Brassicas for dehydrating are chosen frequently for contribution of green colour in a product mix and thus spring cabbage and green leafed cultivars of summer and autumn cabbages are generally used in shredded form. Savoy cabbages perform well as does spinach. Brussels sprouts may be processed whole if the butts are cross-cut to aid drying and rehydration.

7.3.4 *Root vegetables and tubers*

Bulbs have been included later under 'Others', and so dried root vegetables and tubers will include carrots, potatoes, sweet potatoes, parsnips, swedes, beetroot and turnips. Most root vegetables and tubers are dried from cooked diced material and puff drying may be used to give rapid rehydration. Again a typical target moisture content is 5.0%, but lower moisture levels result in longer shelf-lives. Carrots can be cross cut or crinkle cut as rings and dried as such, and starch coating may be used to extend the shelf-life of the dry product because oxidation of carotene together with off-flavour and odour can be a significant problem during storage.

Potatoes can be dehydrated by several methods to give dice, flakes, mashed potato granules and powder, and the reader is referred to detailed publications on this subject (8). Potato crisps, chips and other snack foods are, of course, also dried vegetables.

7.3.5 *Others*

Other vegetables popularly dried include celery, celeriac, leeks, onions, garlic, sweet corn, mushrooms, horseradish and herbs such as parsley, thyme, sage and mint. Asparagus and spring onions are also successfully dried and pimiento and red and green bell peppers may be included.

Freeze drying is particularly useful for delicate vegetables requiring rapid reconstitution properties, and where the on-cost of the process is not so significant, for example, asparagus and mushrooms. In most cases sulphite treatment is used to maintain colour and flavour stability, consistent of course, with relevant legislation as far as residual levels in the end products are concerned. Horseradish relies on enzymic activity to develop its pungent flavour and is not blanched or cooked but handled as delicately as possible until after drying when the enzymes cannot react due to low a_w. The dry material can then be ground, disrupting the cells so that enzyme activity can occur when rehydration takes place, to give full flavour development.

Sliced, flaked or kibbled onion are major ingredients mostly dried on continuous belt conveyors to a target moisture content of 6.0% or less. Onions are not blanched to avoid the flavour loss that would otherwise occur during this process.

Dried root vegetables, potatoes and onions find extensive use in food manufacture both in dry product formulations and as ingredients for wet soups, baby foods and ready meals.

7.3.6 *Storage and packaging*

For bulk storage of dehydrated vegetables, best protection is given by airtight containers flushed with an inert gas such as nitrogen. In temperate climates storage lives of two years or more may be attainable. For sale in smaller bulk or retail sizes packaging with good barrier properties to water vapour, oxygen and light should be used to obtain maximum shelf-life.

7.4 Salt

Salt is present in nearly all pickles and sauces, and in most other processed vegetables. It contributes to flavour, exerts a preservative effect (9), and has certain influences on texture and other characteristics of pickles. It is occasionally added directly as the base for spice concentrates, or is absorbed through its presence in blanching liquor; but in pickles and some sauces, most of the salt present may be residual from the brining or salting of the vegetables used. Brining or salting represents the main use of salt in pickle and sauce manufacture, and it is important that the salt should be of good quality, uninfected with halophilic bacteria, low in calcium, magnesium and iron content, of good white colour and free from foreign matter.

For large-scale production of brine, automatic brine-making installations are available providing saturated brine which may be diluted to the required concentration at the point of use.

The strength of a brine is often determined by the use of a hydrometer. Different types are used, the most common being the specific gravity hydrometer, the Twaddell hydrometer, and different types of 'salinometer' or 'salometer'. Hydrometer methods can be misleading when used on brines which have already been in contact with vegetables because of absorption of soluble solids affecting the density, and due allowance must be made in these circumstances by reference to analytical data, and for any other components of the brine, such as sugar or lactic acid.

7.5 Salting and brining

7.5.1 *Salting*

Many vegetables can be preserved by dry salting alone, for example, runner beans, root vegetables, and marrow, but this is rarely practised now due to the problems of salt removal and consumer acceptability of high salt foods.

7.5.2 *Fermentation brining*

Vegetables received at the brining station have an extensive flora of microorganisms. The growth of the majority of these is inhibited when the vegetables are placed in a brine giving between 8 and 11% equilibrium salt content overall. Certain types, however, are capable of tolerating this salt concentration. These bacteria bring about desirable changes by fermentation. Some homofermentative bacteria convert sugars almost entirely to lactic acid; others, heterofermentative bacteria, produce acid, carbon dioxide and traces of alcohol and acetic acid.

The overall picture in conventional brining is similar to that occurring in the sauerkraut fermentation. The main organisms involved are the Lactobacteriaceae which produce lactic acid, *Acetobacter* which produce CO_2 and H_2, and yeasts which produce CO_2 and alcohol. Other organisms, some desirable and some undesirable, may develop according to the temperature at which the fermentation takes place. Ideally this is between 15 and 20 °C; at lower temperatures the growth of organisms such as *Leuconostoc mesenteroides* is encouraged.

The essential requirements for traditional fermentation brining are:

(a) an initial brine strength which, with the chosen vegetable liquor ratio, gives an equilibrium salt content of 8% or more overall;
(b) sound vegetables at the correct stage of maturity, undamaged and suitably size-graded;
(c) suitable vessels, which may range from brining tanks to the drums in which the brined vegetables are subsequently stored;
(d) the prompt brining of vegetables on arrival at the brining station;
(e) careful weighing of vegetables and brine for each vessel thus ensuring adherence to the predetermined ratio;
(f) the placing of part of the brine in the vessel before adding the vegetables (to assist in ensuring that there are no clumps of vegetables not penetrated by brine);
(g) suitable intermittent mixing, especially during the early stages of brining, to prevent stratification;

(h) careful checks, and topping up with brine where necessary during the fermentation period (which may vary depending on the temperature);
(i) possibly the final draining off of the brine and its replacement by a fresh brine containing up to 1% lactic acid and giving an equilibrium salt content of about 15% overall to ensure the incidental removal of cloudy exudate, dirt, enzyme activity and surface yeasts, and the provision of a sufficiently high salt content and lactic acid content to prevent further microbiological activity.

If a very low strength brine were used undesirable organisms could develop within the vegetables, and experience has shown that a 40° salinometer brine (10% salt) is as low a salt concentration as can safely be used. It is necessary to use a stronger brine in the first place to allow for brine and vegetables to reach equilibrium. Alternatively a careful check may be kept on the brine strength during processing and this adjusted by the addition of salt or strong brine as appropriate.

In the brining and fermentation process, the bacteria use up the natural sugar present in the vegetables to produce lactic acid. They also release carbon dioxide gas which bubbles to the surface and produces an agitating effect. The fermentation stops when there is no further available sugar. Very strong salt solutions, for example, over 17% salt, inhibit fermenting bacterial growth until the brine has been diluted by some means. Unless carefully controlled this can cause problems during freshening, as the sugars will still be available for fermentation.

Providing the container is airtight, the brine strength is not less than 10% salt, and there is at least 0.3% lactic acid in the brine, then, when the fermentation is over, the vegetables can be stored and should stay in good condition for many months.

7.5.3 *Cauliflower brining*

Cauliflower provides a detailed example of a brining operation. Suitable types are summer and autumn heading cauliflowers and the winter hardy spring heading type which becomes available in spring and early summer. The latter is sometimes called broccoli. The pickle manufacturer will require a head of brined cauliflower which is hard in texture, white in colour and with a tight curd.

Harvesting at the right time is essential, and such factors as frost, rain and exposure to sunlight can all cause curd discoloration. It is common practice to break a leaf over an unharvested plant to protect the curd from light, and sometimes frost. It is also important to keep the leaves on the plant for protection from bruising and soiling during transit to the brinery. Then leaves and surplus stalk are removed from the heads, leaving the minimum amount of stalk to allow the heads to hold together. Grading may be carried out at this stage or after the initial brining process. Alternatively, a rough grading can be done to be followed by more selective grading during subsequent processing. Apart from the whole heads, there is some demand for specially cut or diced cauliflower and even for stalk and stump. This can provide an outlet for heads which have been damaged or have suffered partial discoloration or bruising.

Brining can be divided into an initial treatment known as 'shrinking', and brining proper, where full fermentation occurs. The shrinking process consists of putting the cauliflower florets into brine for a short period, during which time, although the volume change is negligible, they become more pliable so that it is then possible to pack more into the same volume container. Shrinkage is normally carried out in suitable vats or drums, the vegetables being held in the shrinking brine for 24 to 48 hours, after which they can be packed into their final containers if desired, covered with brine, and allowed to ferment naturally.

It is recommended that a 14% salt brine should be quite satisfactory for both shrinking and storage. However, it is important to remember that the container will hold somewhere between 60 and 75% cauliflower and the brine will naturally come into equilibrium with this during processing. It is necessary to check brine strength during processing and to adjust it if necessary to maintain it between 10 and 16% during fermentation and subsequent storage. Good agitation is vital during the early stages of brining to avoid stratification. Agitation is usually achieved by rolling the drums at least once a day for the first seven to ten days. If other containers are used it is necessary to employ some form of mechanical agitation to ensure that the brine is thoroughly mixed.

Gaseous fermentation will proceed for around 6 to 8 weeks, depending on the temperature, and during this time some provision for the venting of the container and topping up with brine at regular intervals must be made. The venting provision should restrict the re-entry of air to a minimum. The entry of oxygen will allow the growth of scum yeasts on the brine surface, with a resultant possibility of low acidity and texture deterioration. Lactic acid acidity increases during the process, and at the end of fermentation the brine may be expected to contain around 0.5% lactic acid but can range from 0.3% to as high as 1.0%. In the event of the brine acidity being less than 0.3% it should be adjusted to this figure by the addition of lactic acid, or increasing the sugar level to allow fermentation to progress further. On completion of the fermentation process the vegetable may be repacked. Sulphur dioxide may be added if required, and the containers are finally topped up, sealed, and are ready for storage or use. A completely fresh brine for the final packing may be used if desired containing appropriate levels of salt and lactic acid.

7.5.4 *Cucumbers and gherkins*

Pickled cucumbers represent the main pickle trade in the USA and continental Europe, and an extensive industry and technology has developed in the handling of different types, including brine-fermented cucumbers, cucumbers fermented with dill and spices in a mildly acid brine, and pasteurised fresh cucumbers.

In the UK, the main interest is in small cucumbers or gherkins. Size graded brine-fermented gherkins are usually imported from Southern Europe and Holland. Size is normally expressed in terms of number per kilogram, and each grade is quoted as a range of sizes, for example, 30/40; 80/100; 180/200. The sizes larger than 50s are normally used for cutting into rings, dice or spears, for use in mixed pickle, sweet pickle and piccalilli, as are misshapen gherkins of any size.

Gherkins should be brined very promptly after harvesting, because delays result in progressive quality deterioration. Where delays between harvesting and brining are unavoidable, it is possible to minimise deterioration by controlled atmosphere refrigerated storage. The fermentation brining of cucumbers is subject to the general considerations listed above, but it also involves the peculiar problem of 'bloaters' and 'floaters', caused by the presence of cavities in some fruits. Apart from the undesirability of this defect in itself, bloaters are liable to emerge from the brine and develop mould. Factors influencing the incidence of bloaters include cultivar, size, the use of nitrogenous fertilisers, salt concentration, the addition of sugar or lactic acid to the brine, yeast contamination of brine, and high brining temperature. Gas production is due to gas-forming microorganisms occurring within the tissues. Pricking is sometimes practised, primarily to assist in securing rapid brine penetration, but it is also beneficial in allowing release of internally formed gas, and thus in minimising incidence of bloaters. The addition of 500 ppm sodium benzoate to the brine (where permitted) also minimises the problem.

7.5.5 *Onions*

The two main types of pickled onion packs are silverskin onions and what are usually referred to as 'picklers' or 'brown onions'.

Nearly all silverskin onions are grown in Holland, with Israel a smaller supplier. Cultivation is interesting as the bulbs need to remain below the soil surface to prevent exposure to light and greening. The leaf is burned or mown off just prior to lifting, and a soil layer complete with onions is collected up by the harvester and the bulbs and soil separated by dropping the soil back to the field. Onions are fully fermented in brine, or delivered fresh chilled or in acid liquor for immediate processing. When brined, they are graded by size (for example, 18–21 mm), colour, and shape, but only for size and major defects if fresh or in acid liquor. Sulphur dioxide is usually present in the brine or the acid liquor. Fermented silverskin onions should have a white, translucent appearance, crisp texture and characteristic flavour.

At one time brown pickling onions were fully fermented in brine but changing tastes and economics have meant that now virtually all are quick-brined, that is, semi-preserved in weak brine. Onions used to be mainly supplied as outgrades from ware crops, but now specific cultivars are available to produce small onions suitable for pickling. Most supplies may be obtained ready-peeled from specialist vegetable preparers.

In the quick-brined process the onions do not ferment significantly and are only in brine for a short time, thus retaining much of the fresh onion flavour. The onions are brined for at least 24 hours to prevent cloud formation in the jar, but with a low salt content to save freshening, and a high lactic acid content in the brine to retard fermentation by the repression of microbial growth. The most probable causes of problems in this process are excessive delays between peeling and getting the onions into the brine, and failure to ensure that the onions are properly submerged below the brine surface. Ideally, there should be at least two clear inches between the uppermost onions and the brine surface.

7.5.6 *Red cabbage*

Red cabbage is most frequently used in the fresh state, possibly being subjected to a brief brining after shredding. It must be handled on the same day as it is received at the factory.

7.5.7 *Walnuts*

Walnuts are harvested in late summer while under-ripe and before the shell has hardened. Once the woody shell has formed, it cannot be softened. In order to facilitate penetration of brine through the tough outer tissue, the nuts can be punctured. Brine fermentation follows the normal pattern.

7.5.8 *Other vegetables*

Similar brining techniques to those described above can be used with other vegetables, and many involve the use of other acids than lactic with or without the additional use of permitted preservatives and processing techniques such as blanching.

7.5.9 *Sauerkraut and the sauerkraut fermentation*

Sauerkraut is produced from hard white or red cabbage varieties. After trimming and fine shredding, the cabbage is mixed thoroughly with dry salt at a level of 2.0–2.5% and pressed down in tanks under weighted covers. Sufficient brine to cover the cabbage is formed by the release of tissue fluid from the shredded cabbage dissolving the salt. A lactic fermentation ensues, taking some three to six weeks, and giving a lactic acidity of 1.8–2.2% and a pH of 3.5–3.7. A typical course of fermentation is illustrated in Figure 7.4. The microflora involved are normally naturally associated with the cabbage but

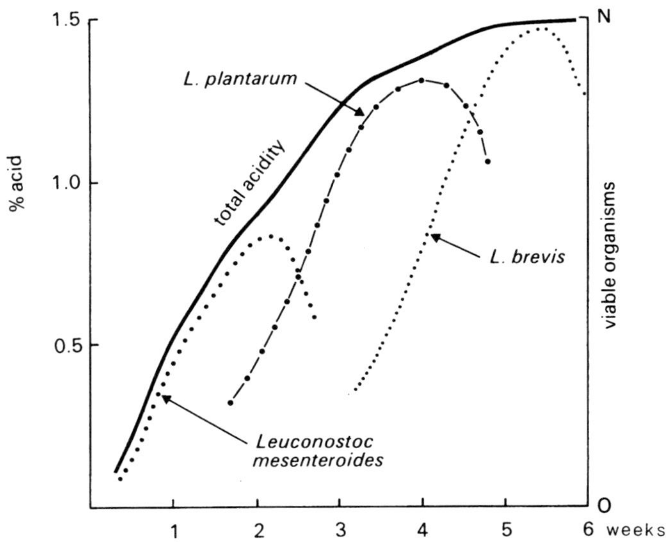

Figure 7.4 Typical sauerkraut fermentation showing cumulative acidity and microbial population.

starter cultures can be used. Temperature control during fermentation is critical.

At the end of fermentation the sauerkraut is packed into suitable containers and pasteurised, possibly with the addition of other ingredients to market taste. The nutritional and economic contribution of sauerkraut is worthy of mention as it is a valuable source of vitamin C, and is among the highest per capita consumption for processed vegetables in several parts of the world.

7.5.10 *Olives*

Olives are not freshened, but are packed in brine, and are usually considered, to be pickles if not actually acetic-acid preserves. To produce the end products, the vegetables are packed from fermented stock, either in original brine, or repacked in a brine containing 8–10% salt and 0.5% lactic acid. They are normally hot-filled or pasteurised to prevent scum yeast formation, but spoilage may still occur after opening. Oil is sometimes added to avoid this, and preservatives may be considered. The brining of olives is interesting in that the glucoside oleopein, which makes them very bitter when fresh, is hydrolysed by a lye treatment before brining, which reduces the bitterness. Olives may be pitted and/or stuffed after brining.

7.5.11 *Fermentation inhibition in brining*

The onset of fermentation may be delayed or inhibited by a variety of techniques. These include:

- The use of high acid brines (lactic and/or acetic)
- The use of preservatives, for example, sulphur dioxide and sorbic acid
- Chilling
- Blanching, coupled with
- Flash pasteurisation of liquor and aseptic filling
- Bulk pasteurisation
- Combinations of the above.

The advantages of retarding or preventing fermentation are those of cost, convenience and, depending on the case in point, quality.

7.6 Freshening or debrining

Before brined vegetables can be used, their salt content has to be reduced to an acceptable level for the eventual consumer, usually below 5% in the freshened vegetables. This is done by a reverse of the brining process, that is, washing the salt out with water (11). Stratification can be a problem, because salt coming out of the vegetables will tend to sink to the bottom of the container. This can be avoided by some means of agitation, for example, rolling the drums between changes of water or using a continuous flow of water so that all the salt is washed away quickly. The use of freshening water that contains a high level of calcium can improve the texture of the vegetable in the end product.

It is important to use potable water, to avoid contamination with iron and copper, and to ensure that freshened vegetables are used within a reasonable period of time. Freshened vegetables may be acetified pending use, to ensure ready availability of stable desalted stock.

Many factors affect the rate of removal of salt from vegetables, including the type and size of vegetable, temperature, and the concentration gradient of salt between the outer part of the vegetables and the liquid immediately in contact with it. The last is affected by other factors such as relative quantities of vegetables and water, flow rate or frequency of changing of the water, and agitation.

Methods of freshening fall into the main categories; continuous, multibatchwise and single batchwise.

7.6.1 *Continuous methods*

The crudest continuous method is to drain the brine off, put a hose into the cask or drum and leave the water running. This is unlikely to give satisfactory or uniform results because the denser brine tends to fall. The correct technique for continuous debrining is to introduce water at the top of the cask or drum and run it to waste at the bottom.

Continuous methods are the most rapid and require the least labour in handling, but are normally rather wasteful of water. The optimum results may, however, be obtained by combining a slow water flow with agitation. The ideal system consists of a large stainless steel or plastic tank, not more than a metre deep, and capable of holding the vegetables from several drums of vegetables. It has a water inlet valve at the top and an outlet valve at the bottom a few centimetres above the base of the tank. The tank is fitted with a means of agitation or circulating the water, and the rates of outflow and inflow of water are balanced to maintain a constant level.

The rate of debrining will depend on the precise factors of size and shape of tank, stirring conditions, type and quantity of vegetables (onions lose salt more slowly than gherkins or cauliflower), and flow rates. Initial experiments with the equipment available will establish a standard procedure. Once this has been done for each type of vegetable, one can operate under the prescribed conditions for the standard debrining times thus established, merely checking the salt content at or near the end of the debrining period. After debrining, acidification of the vegetables may be carried out in the same tank, with agitation but with valves closed.

7.6.2 *Multibatchwise methods*

Multibatchwise methods involve repeated changes of water, and the speed of debrining depends on the frequency of the changes, and the extent of any agitation. Without agitation, this method inevitably gives rise to stratification. Thus the multibatchwise system without agitation must be regarded as unsatisfactory.

7.6.3 *Single-batchwise method*

The single-batchwise method, involving the suspension of the vegetables in a stainless steel wire-mesh basket in the top part of a vessel of water, turns to advantage the otherwise disadvantageous occurrence of stratification. Thus, the brine formed by extracted salt falls into the lower part of the vessel, and at no stage is there a significant variation of salt content between the vegetables at the top and bottom of the basket. In this method, agitation must be avoided. Again the precise vegetable to water ratios and times to be used will need to be determined by experiment and analysis for each size and type of vegetable.

7.7 Pickling

In general terms products with an expected pH of greater than 4.5 have been excluded from this heading. Literature on pickles and pickling is not extensive

(12, 13, 14). In most cases the principal acid involved will be acetic, normally derived from the vinegar ingredient, but in some cases it will be lactic acid from fermentation processes, as with olives, citric acid from lemon or other fruits, or malic acid from apples. In most situations a number of factors will contribute to the total preservation system, product integrity and stability, for example, salt content, soluble solids and water activity, any thermal processing treatment given, and the presence of any preservatives and antioxidants. In almost any products of this type a wide range of recipes can be possible, reflecting different approaches, different ideas of what will appeal to consumer tastes, ethnic and local specialities, different choices of methods and equipment, and varying economic considerations. Choice in these matters can only be exercised by the producer within the constraints of the law and good manufacturing practice.

One great benefit of products of this nature is that there is minimal risk of food poisoning from them so long as there is some acetic acid (about 0.5%) present, and even where acetic acid or vinegar is not used, as with a mayonnaise made only with lemon juice for example, so long as a low enough pH is achieved pathogenic microorganisms will not grow.

7.7.1 *The preservative effect of acetic acid*

The acetic acid present in pickles, chutneys and sauces is the main factor responsible for the self-preservation of these products, coupled to a lesser extent with other preservative influences. Unpasteurised products containing no added preservatives rely on the acetic acid present only, while pasteurised products may further rely on its combined effect with heat processing, at least until the container is opened.

In acetic acid preserves, the action of acetic acid is bacteriostatic, and not directly dependent on pH. Its inhibitory effect is due to the undissociated acetic acid molecules which are thought to have the ability to penetrate the plasma membrane and act as a proton ionophore. pH is involved, however, for the degree of dissociation of the acid depends on the pH of the product in which it is incorporated. Fortunately, at the pH generally obtaining in pickles, nearly all the acetic acid is in the undissociated form.

In evaluating microbiological keeping quality in relation to product composition, the acetic acid present calculated as a percentage of the whole product is not a reliable criterion, but rather the acetic acid acidity calculated as a percentage of the total volatile constituents:

$$\frac{\text{Total acetic acidity} \times 100}{(100 - \text{Total solids})}$$

a value referred to as the Preservation Index (PI).

If the PI is greater than 3.6, a reasonable degree of freedom from microbiological spoilage can be expected, although the relationship is arrived at empirically and is subject to some important constraints. Certain lactobacilli and yeasts can tolerate acid at a PI of up to 3.6, and a mould has been encountered tolerating and metabolising acetic acid at even higher levels. The relationship takes care of the influence of soluble and insoluble solids, including salt and sugar within common ranges. At high levels sugar and salt probably have a preservative influence in their own rights which can permit a reduced PI.

Equilibrium within the tissue of the different ingredients takes some time to achieve, but fortunately spoilage organisms are not normally to be found deep within the tissues, so for quality control purposes it is possible to take into account the anticipated degree of equilibrium which will have been reached when analysing covering liquors. These yardsticks presuppose the application of reasonable standards of hygiene.

Acetic acid preserves with acidities below the 3.6 index are safeguarded by adequate pasteurisation or other equivalent regime of heat treatment and sometimes also by chilled storage.

7.7.2 *The influence of other acids*

The influence of other food acids is primarily related to their effect on pH, and to a small extent therefore on the buffering characteristics of the whole system. None has the preservative capability of acetic, although all have characteristic flavour contributions to make to the final product.

7.7.3 *Raw materials*

7.7.3.1 *Fruits.* The terms 'fruit' and 'vegetable' in the pickle trade do not correspond to a strict botanical classification. Tomatoes, cucumbers, gherkins and walnuts are normally considered in the trade as 'vegetables' and therefore are dealt with under that heading.

7.7.3.2 *Tomato purée.* Tomato paste or purée (15) is an important ingredient used in the manufacture of tomato ketchup, relishes, other sauces, and sweet pickle. For tomato chutneys and relishes, part or all of the tomato content can come from whole tomatoes. Apart from this, canned concentrated tomato purée is ideal for supplying the required tomato content, being prepared when tomatoes are in season and therefore at their cheapest, available for use all the year round, and usable without further preparation. The main producers of tomato purée are countries in southern Europe, the eastern Mediterranean, the USA (notably California) and North Africa.

Canned tomato purée may be in the form of 'single concentrate' (18–20% total solids), but more usually as 'double concentrate (28–30% total solids) or 'triple concentrate' (38–40% total solids). These traditional descriptive terms are misleading, and represent approximately fourfold, sixfold and eightfold

Table 7.1 Vegetables for the manufacture of pickles.

Vegetable	Fresh/ chilled	Semi-preserved	Brined
Onions and silverskin onions	+	+	
Cauliflower	+		+
Gherkins and cucumbers	+	+	+
Walnuts			+
Olives			+
Capers			+
Red cabbage	+		
Beetroot	+	+	
Marrow	+	+	+
Carrot	+	+	+
Rutabaga or swede	+	+	+

concentration respectively of the original pulp. Most of the tomato purée imported into the UK is double concentrate. Although triple concentrate has always been the more economical way of obtaining a given amount of tomato solids this can be at the expense of colour, flavour and texture. However, triple concentrate tomato purées can be found with satisfactory colour and flavour characteristics.

The standard pack for industrial tomato purée is the '5 kilo' can. The '5 kilo' designation is rather misleading, as the actual net weight can be significantly less than 5 kg. Bulk packs filled aseptically are increasing in use.

7.7.3.3 *Vegetables.* Vegetables used in pickle manufacture may be supplied fresh or chilled, semi-preserved, or fully brined, typically as indicated in Table 7.1.

7.7.3.4 *Fresh and chilled vegetables.* With the exception of gherkins, silverskin onions, capers, olives and walnuts, most vegetables are now grown extensively in the UK. Cauliflower for mixed pickle, piccalilli and chutneys are normally brined, but they may be delivered fresh to the manufacturer. Both the true cauliflower and winter cauliflower (formerly known as broccoli) are used for brining and pickling.

Most pickling red cabbage is used fresh taking good coloured cultivars with tight heads such as Langedijk. Marrow can be useful as an alternative to the more costly cucumber or gherkin as a filler vegetable in sauce pickles. Root vegetables such as carrot and swede may be used fresh and incorporated into sauce pickle and chutney formulations, without further processing other than peeling and cooking.

Many other fruits and vegetables are used, according to price and availability, throughout the world, for example, mangoes, apples, pears, maize, and water chestnuts.

7.7.3.5 *Semi-preserved vegetables.* These may be received in a covering liquor or may have been recently drained of one. This is not of itself sufficient to ensure long-term stability during storage, but it will suppress or inhibit short-term deterioration, for example, gherkins in weak acid solution and onions in weak acidified brine.

7.7.3.6 *Fully brined vegetables.* Fully brined vegetables may be regarded as stable, subject to reasonable safeguards during storage. They are either fully fermented products or in a brine so treated as to inhibit the onset of fermentation completely. They may be stored for periods of months or even years before use and may have been subjected to considerable preparatory processing, for example, walnuts and olives.

7.7.3.7 *Vinegars.* Vinegars are the product of two successive fermentations (without intermediate distillation, except in the case of spirit vinegar), whereby a yeast fermentation converts sugars to alcohol and an aerobic *Acetobacter* fermentation converts the alcohol to acetic acid. Other minor reactions occur to give the characteristic flavours of vinegars. Any raw material containing sugar, or starch which can be converted enzymically into sugars, can provide the feedstock.

Acetic fermentation can be achieved in quite simple traditional trickling 'acetators' comprising a support material, such as birch twigs, in a column allowing good airflow while alcoholic liquor is trickled over it and recycled, to modern sophisticated acetators operating on the principle of forced aeration dispersing bubbles through the liquor and maintaining constant agitation.

Malt vinegar is prepared from malted barley, often with other cereals, ground and steeped in water to produce a 'mash' in which the malt enzymes convert starch to soluble sugars and dextrins. The 'wort' resulting from this is drained off and inoculated with yeast to produce an alcoholic liquor or 'gyle'. After storage, the matured gyle is subjected to oxidative fermentation by *Acetobacter*.

Distilled vinegar is simply any vinegar prepared by distilling off the acetic acid and other volatile constituents. It is white in colour and because it is free of many impurities is much less likely to cause discoloration in the end product in which it is used.

Spirit vinegar is a brewed vinegar, usually prepared from the distilled alcohol from the fermentation of molasses, and is available economically at high strength (upwards of 10%) and virtually colour free. Hence it is widely used as a manufacturing ingredient.

Wine vinegars are, of course, fermentation acetified wine, hence the derivation of the name vinegar – *vin aigre* – sour wine.

Other vinegars arise from conversion of virtually any alcoholic liquor; thus cider vinegar, honey vinegar and citrus vinegar are all to be found. There are even equivalents of the alcohol industries fine wines, such as the Balsamic Vinegar from Modena, produced from high-quality grapes and aged in oak and mulberry. In addition there is a market for flavoured vinegars such as tarragon, dill, and garlic, and for spiced vinegar for home pickling.

7.7.3.8 *Others*

Acetic acid is the acid constituent of vinegars, and has powerful properties of microbial preservation. It is commercially available at around 80% concentration for food manufacture and is transported by bulk tanker or in drums. Suitable conditions of storage and appropriate health and safety precautions are required. Acetic acid may be used wholly or partly in place of vinegar as a food ingredient, as a preservative acid for work in progress, or simply as an acidity regulator for fine adjustment of final acid levels.

Lactic acid is a colourless, odourless, non-volatile acid produced by a fermentation process and normally concentrated to 50 or 80% for food purposes. It is naturally produced in the brine fermentation of vegetables and is important as a pH reducer in this context. It is added to make up natural shortfalls or losses for this reason, and makes a flavour contribution in its own right. In the end product it is said to lessen the sharpness of the acetic acid flavour.

Water is an important item as an ingredient, as a heating and cooling medium, and for washing and cleaning ingredients, packaging, plant and equipment. For all purposes a good potable quality is required, and as an ingredient it should be non-alkaline and free from traces of hydrogen sulphide and iron.

In many cases calcium can be advantageous to vegetable texture, and hard water for ingredient use is often desirable. However, significant calcium or magnesium hardness can be undesirable, and water should be soft or softened for general production purposes, and for use in beetroot formulations.

7.8 Pasteurisation of acid products

Pasteurisation (10) of pickles involves the controlled and limited heat-processing of the final product within its container, with the following advantages.

(1) The heat destruction of highly acetic acid-tolerant spoilage microorganisms, thus preserving the pack during its shelf-life, the acid present continuing to preserve the product thereafter.
(2) Total or partial heat-inactivation of enzymes of vegetable or of

microbiological origin, thus preventing or inhibiting deterioration reactions caused by enzymes.

(3) In cases where self-venting closures are used, the removal of air during pasteurising and retention of headspace vacuum on cooling will minimise oxygen-dependent and oxidative deterioration reactions, including darkening and softening.

Pasteurisation of pickles in the USA was practised from the late 1930s, and process definition in terms of in-pack temperature and time requirements were related to microbial thermal death-times. More recently, aspects of thermal inactivation of enzymes implicated in softening have been considered.

In general terms, the process required for the inactivation of potential softening enzymes is more severe than that needed for destruction of acetic acid-tolerant organisms, and hence, it may be beneficial to pasteurise even high-acid products for this reason alone. Also, where high pasteurisation temperatures are used this can contribute further cooking effect to some products if so desired. Thus, pasteurisation of lower-acid pickles is ideally suited to the production of crisp, mild vinegar pickles of good appearance and stability, and may also benefit other types.

Pasteurisation of filled capped jars may be achieved in a variety of ways, including batch methods, in which a batch of jars is heated in a steam cabinet or a tank of hot water, and continuous methods, in which jars are conveyed through a steam chamber, a hot-water spray chamber, or a hot-water bath. Pasteurising conditions depend on;

(a) the temperature–time requirement at the point of slowest heating in the jar (in most packs two-thirds to four-fifths of the way down the central axis);
(b) the time needed to raise the contents at that point to the required temperature, which in turn depends on the rate of heat penetration through the glass and contents; and
(c) the initial temperature.

Reference should be made to the literature for a detailed account and discussion of the work that has been done on pasteurisation (10, 16). Typical pasteurisation regimes are given in Table 7.2.

Table 7.2 Typical pasteurisation regimes

Product	Temperature (°C)	Temperature (°F)	Time (min)
Onions	79	175	10
Silverskin onions	71	160	15
Gherkins/cucumbers	74	165	20–25
Mixed pickle	71	160	15
Beetroot	82	180	20

The processing conditions which will be required will depend on the nature of the pasteurising equipment and method of operation, the size and shape of the jars or bottles, the precise composition of liquor, the vegetable/liquor ratio, and for a given set of circumstances, processing conditions *must* be established experimentally.

Process determination in continuous pasteurisers is more difficult than for batchwise, but the greater uniformity of processing attainable has led to wider use of P-value calculations for process evaluation and definition (22). Using the principles described by Shapton, Lovelock and Laurita-Longo (16) and Stumbo (17), and presuming $z = 7\,°C$, for most purposes a minimum process value of $P_{70} = 10$ can be considered.

Measurement of in-pack temperatures is difficult in continuous pasteurisers, and it is essential that measurements are made that include significant parts of the warming-up and cooling-down zones. A travelling recorder may be used for this purpose. For products likely to suffer from overcooking it is essential to achieve the shortest warming and cooling times consistent with good glass-handling practice. In some cases, particularly very large packs, a compromise will need to be made between enzyme inactivation effects and adverse cooking effects.

Excessive headspaces in jars of pasteurised pickles may result not only in poor appearance but in failure to comply with the declared average weight of contents. This may arise from trapped air (especially in beetroot), interstitial air in the vegetables, or the presence of dissolved carbon dioxide driven off and expanding during pasteurising and either rising to the headspace or actually driving out liquor (with possible continuance of liquor loss by capillary action).

Pasteurised low-acid pickles have a PI of less than the recommended minimum of 3.6%. When these pickles are opened in the home, however, they do not usually undergo microbiological spoilage, the probable reason being that as acid-tolerant pickle spoilage organisms occur largely in pickle factories, they are not generally distributed, and are unlikely to be found in the home. Those microorganisms which have been found in the home can be inhibited by the lower level of acetic acid present in the pickles. Pasteurisation of low pH products and the potential problems associated is reviewed by Anderson and Campbell-Platt (13).

7.9 Preservatives

The presence of preservatives in pickles and sauces will be governed by national legislation, but these may permit them to contain:

up to 100 mg/kg sulphur dioxide,
250 mg/kg of benzoic acid or of methyl- or ethyl- or propyl-4-hydroxybenzoate
1000 mg/kg of sorbic acid.

Vinegar may contain sulphur dioxide. The appropriate label declaration must be made, and as with all food ingredients, the current local legislative situation should be checked by the intending user prior to final commitment.

7.9.1 *Sulphur dioxide*

Food regulations in many countries permit the use of sulphur dioxide up to a level of 100 mg/kg in pickles and sauces, and also allow its use in the ingredients provided that the necessary declaration is made. Hence higher levels than 100 mg/kg can be used in the brined stock or fruit or vegetable pulps if wished so long as the maximum allowed in the end product is not exceeded.

Sulphur dioxide (100 mg/kg) inhibits the growth of pickle spoilage lactobacilli, yeasts and moulds in media in the presence of acetic acid, but some of the yeasts and moulds grow readily in the presence of 50 ppm. However, only uncombined sulphur dioxide is inhibitory, and because substantial proportions of the sulphur dioxide in a pickle or sauce rapidly combine with components of the product, at 100 mg/kg it is not an effective preservative from a microbiological viewpoint, with the exception of olives, where it has been shown that it effectively prevents the formation of scum yeasts.

The benefits deriving from the use of 100 mg/kg sulphur dioxide spring from its reducing or antioxidant properties, and its property of inhibiting the Maillard non-enzymic browning reaction. Thus, it minimises storage deterioration of colour, odour and flavour in gherkins; retards darkening of onions, with improved retention of satisfactory odour and flavour; confers white colour, minimises storage deterioration of odour and flavour in silverskin onions; gives whiter cauliflower; reduces discoloration of cauliflower which has been coloured with turmeric in mixed pickle; greatly minimises loss of crispness and dulling of colour in red cabbage; prevents blackening by trace iron contamination; minimises surface darkening in non-mayonnaise based horseradish relish; and prevents (but will not reverse) 'pinking' of onions.

The main adverse effects are in salad cream, and products based on it such as sandwich spread and horseradish cream, where addition of sulphur dioxide can result in emulsion breakdown and development of off-flavours; and in pickled beetroot, where it results in marked colour loss.

In brined vegetables sulphur dioxide is of value in preserving the white colour of cauliflower and silverskin onions, although the time of addition is important and it is the author's view that it should be added only after the fermentation is complete, as earlier addition may retard fermentation to an undesirable extent. It may also mask defects such as iron contamination which will come to light during the pickle processing.

7.9.2 *Benzoic acid and the alkyl-4-hydroxybenzoates (parahydroxybenzoates)*

In general, the hydroxybenzoates have not proved particularly effective, even at the maximum permissible levels, against pickle spoilage microorganisms. Benzoic acid itself has not been significantly popular, except where it is used in cooked and pre-packed beetroot in which it is also specifically permitted in the UK.

7.9.3 *Sorbic acid*

Sorbic acid is a useful permitted preservative, shown by Dakin (25) to be effective against yeasts in brined vegetables, *Lactobacilli*, and is particularly effective against the mould *Moniliella acetoabutans*.

7.10 Syruping

Syruping is a means of reducing water activity to improve microbiological stability, and while used for cherries and other fruits is rarely applied to vegetables. However, it can be used as a technique for preparation of vegetables for inclusion in sweet pickle products to prevent flotation of vegetables in sweet sauces and local dilution of sauce by diffusion of tissue fluid from vegetable ingredients.

7.11 Hurdle and combination techniques

Combination or hurdle techniques are those where two or more preservative effects combine synergistically to moderate the requirements or effects of an individual treatment. Examples have been described above wherein salt, acid and preservative effects combine to allow, for instance, a reduced thermal process or an increased storage time. Other examples might include, for example, a blanching treatment coupled with controlled atmosphere packaging.

Irradiation treatment has been demonstrated to provide benefits in different vegetables, for example, suppression of sprouting in potatoes and onions, extension of shelf-life of prepared carrots, and it is likely to establish itself as a useful technique in this area. It has been used for some time in certain countries for reducing or controlling microbial load and infestation in herbs and spices, and in South Africa to reduce incidence of mould growth in tea in areas of high humidity.

Where hurdle techniques are to be considered it cannot be emphasised too strongly that the intended regime must be subjected to critical scrutiny by appropriately qualified persons to ensure that freedom from microbiological hazard is unimpaired. The use of hazard analysis critical control point (HACCP) techniques in this regard are to be commended.

7.12 Colours

It is not generally necessary to add colour to vegetable products. The natural red colour of tomato sauce or ketchup, pickled beetroot and pickled red cabbage, made from good-quality ingredients, is carried through to the product.

Yellow and green colours are those most commonly used. Turmeric contributes a characteristic flavour and yellow colour, but the yellow is unstable to light, fading rapidly in sunlight and under fluorescent lighting. Tartrazine and quinoline yellow can be used, and also the natural colours annato, lutein and riboflavin. For green coloration, mainly for green beans, peas and mint products, chlorophylls may be used, or mixtures of coal tar colours.

Any colour that is added must:

- Be reasonably stable when treated under the conditions used in manufacture;
- Be stable in the end product in both light and dark, and at low pH in the presence of vinegar and low salt concentrations;
- Be permitted for use in the product in the country in which it is to be sold;
- Be declared in accordance with relevant legislation.

Caramel is of particular interest to the pickle maker, giving a dark brown colour to malt vinegar, brown sauces, and relishes. Different caramels have varying isoelectric points, that is, the pH value at which the colloidal particles are electrically neutral. Above this pH value the particles are negatively charged and below it they are positively charged. When positively and negatively charged colloidal particles come together they coalesce and precipitate. Malt vinegar contains some positively charged colloidal matter especially if it is not fully matured. If caramel is added, some precipitation may occur, so the caramel addition should be made before the final filtration which clarifies the product. By contrast, pickled onion cells are negatively charged, so any added caramel colloidal particles must also be negatively charged at the pH of the liquor (usually 2.5–3.0) to prevent precipitation.

7.13 Enzymes

The most important enzymes for pickles and other pasteurised vegetable products are those responsible for softening. They are mainly those causing degradation of cell wall pectin which is partially methoxylated polygalacturonic acid. The pectinesterase group of enzymes attack and split off the methoxyl side groups, while the polygalacturonases attack the links between the galacturonic acid units making up the chain. Moulds are a primary source of these enzymes.

The rate at which enzyme reactions proceed is temperature dependent, and there is much to be said, therefore, for storing stocks of vegetables in brine in the coolest possible conditions. Enzymes can be inactivated by heat, and the blanching of vegetables and the pasteurisation of pickles serves to prevent or minimise enzymic deterioration as well as to help to prevent microbial spoilage.

7.14 Packaging

The packaging requirements for these products are like those of any other foods, and the reader is referred to Chapter 8 in this volume and the publications of Paine and Paine (20) on general matters, and that of Moody (21) on packaging in glass. Again, prime considerations relate to corrosion resistance to acetic and other acids, salt, and sulphur dioxide, impermeability to water vapour, oxygen and sulphur dioxide; closure reseal performance; and practicality in use.

7.15 Spoilage of acid products

The main kinds of spoilage of pickles and sauces are:

- Multiplication of yeats, moulds or bacteria;
- Deterioration of appearance, texture or flavour through the action of enzymes of vegetable or microbiological origin;
- Deterioration of appearance, texture or flavour through oxidative reactions;
- Deterioration due to the action of trace metals;
- Deterioration (mainly of appearance) due to the physical or chemical interaction of component substances;
- The presence of foreign matter or contaminants of any kind;
- Deterioration due to unsatisfactory packaging or adverse storage conditions.

In many real instances of spoilage two or more of these basic types may be involved simultaneously, for example, oxidative rancidity in salad cream accelerated by trace metal contamination; or discoloration due to the reaction of a component with a substance resulting from an enzymic reaction; or the influence of storage temperature on any of the first five items.

Many of these kinds of spoilage are referred to elsewhere in this chapter, under appropriate headings, but a summary is given below.

7.15.1 *Multiplication of yeasts, moulds or bacteria*

Because of the presence of undissociated acetic acid and their low pH, pickles and sauces present a very low risk of microbiological food poisoning. There is, however, a possible risk of the formation of staphylococcal toxin in incorrectly handled ingredients with subsequent heating being inadequate to destroy any toxin which might be formed.

Leaving aside the question of chemical preservatives, the problem of microbiological spoilage is largely one of microorganisms capable of tolerating acetic acid, and can be considered for three concentration ranges of this acid;

(i) Up to 1.0% acetic acid in the volatile constituents. Many organisms can grow at low levels of acetic acid, depending upon the medium, its pH and the presence of other preservatives.

(ii) From 1.0 to 3.6% acetic acid in the volatile constituents. At these levels only acetic acid tolerant organisms can grow. These are various strains of *Lactobacilli*, and certain yeasts.

(iii) Above 3.6% acid in the volatile constituents. Apart from some acetous fermenters (which tend to lack salt tolerance), the only significant organism is a mould, *Moniliella acetoabutans*. This organism is unique in its ability to tolerate and metabolise acetic acid.

Apart from pasteurisation and preservatives (referred to earlier), the only real safeguard against this type of spoilage is strict hygiene.

With the advent of salad products with low acetic acid levels and high pH values, microbial spoilage hazards must be assessed very carefully. The microbial stability of mayonnaise and cottage cheese based salads was investigated by Dennis and Rose (23). As well as organisms of the type described above, other *Lactobacilli*, *Candida* and *Saccharomyces* have been identified as potential spoilage organisms.

Spoilage by a wider range of microorganisms may occur at intermediate stages of handling and processing ingredients, and subsequent incorporation may spoil the finished product. For example, prolonged debrining of vegetables with poor hygiene and inadequate acidification, may result in fermentation by any of a number of common yeasts. Another example arises from inadequate lactic acid in fermented brined vegetables allowing the growth of otherwise inhibited proteolytic bacteria, elaborating enzymes capable of breaking down proteins.

7.15.2 *Deterioration through oxidative reactions*

Examples of oxidative deterioration are the softening of pickled red cabbage. and the browning of tomato ketchup.

7.15.3 *Deterioration due to action of trace metals*

Apart from the role of trace metals, especially copper, in accelerating oxidative rancidity, and the role of trace iron in the sulphide blackening of vegetables, the main form of trace metal deterioration is discoloration due to trace amounts of iron with tannins.

7.15.4 *Deterioration due to interaction of component substances*

Interaction of components is exemplified by Maillard browning reactions, the 'pinking' of onions and the deposition of caramel on vegetable surfaces.

7.15.5 *The presence of foreign matter or contaminants*

The problem of avoiding foreign matter in foodstuffs has been referred to elsewhere. The avoidance of carry-overs and contaminants which might cause taints or off-flavours is more difficult, and the advantage of using known and reliable supply sources is great in this context.

In general, the measures required are those applicable to all food manufacture and may be summarised as:

(a) Measures to inspect and decontaminate raw materials;
(b) Measures to prevent the entry of foreign matter during processing;
(c) Measures to treat the finished product before filling;
(d) Visual and instrumental inspection of the filled containers;
(e) Organoleptic checking of end products and work-in-progress against off-flavours and taints;
(f) Vigilance on the part of all operatives, and strict supervision.

7.16 Shelf-life

Several aspects relating to shelf-life have already been covered above. Many factors will influence the product shelf-life, such as

- Initial raw material quality;
- Seasonal variation and normal biological variation;
- Packaging integrity and oxygen availability;
- Storage temperature;
- Time;
- Process treatment and its adequacy;
- Enzyme activity;
- Presence of contaminants, e.g. copper;
- Use and level of additives;
- Inherent autolysis;
- Consumer expection;
- Exposure to light.

All of these factors are of significance, but are mostly within the control of the manufacturer. Assuming general adequate control and packaging integrity, as the rates of progressive deterioration reactions are temperature dependent, it follows that the optimum shelf-life will be obtained under the coolest practicable storage conditions, throughout the distribution chain.

Acknowledgement

The writer thanks Brooke Bond Foods Ltd for permission to publish this chapter.

References and further reading

1. Luh, B.S. and Woodroof, J.G., *Commerical Vegetable Processing*, AVI, Westport, CT, 1975.
2. Duckworth, R.B., *Fruit and Vegetables*, Pergamon Press, London, 1966.
3. Desrosier, N.W., *The Technology of Food Preservation* (3rd edn), AVI, Westport, CT, 1970.
4. Earle, R.L., *Unit Operations in Food Processing*, Pergamon Press, New York, 1969.
5. Dalgleish, J.McN., in *Food Industries Manual* (22nd ed), Blackie, Glasgow, 1988, Chap. 14.
6. van Arsdel, W.B., Copley, M.J. and Morgan, A.I., *Food Dehydration*, Vols 1 and 2 (2nd edn), AVI, Westport, CT, 1973.
7. Seltze, E. and J.T. Settlemeyer, *Spray Drying of Foods, Advances in Food Research*, Vol II, Academic Press, New York, 1949.
8. Talburt, W.F. and Smith, O., *Potato Processing* (3rd edn), AVI, Westport, 1975.
9. Ingram, M. and Kitchell, A.G., 'Salt as a perservative for foods', *J. Food Technol.* **2** (1967), 1.
10. Dakin, J.C., *'Pasteurisation of acetic acid preserves'*, in *Recent Advances in Food Science*, Vol II, Butterworths, London, 1962, p. 128.
11. Dakin, J.C., *Freshening of Brined Vegetables*, BFMIRA Research Reports No. 107. Leatherhead, 1961.
12. Binstead, R., Devey, J.D. and Dakin, J.C., *Pickle and Sauce Making* (3rd edn), Food Trade Press, London, 1971.
13. Campbell-Platt, G. and Anderson, K.G., in *Food Industries Manual* (22nd edn), Blackie, Glasgow, 1988, Chap. 8.
14. Anderson, K.G., *The Technology of Clear Pickles*, IFST Mastership in Food Control Dissertation, Serial No. 34, 1981.
15. Goose, P.G. and Binstead, R., *Tomato Paste and other Tomato Products* (2nd edn), Food Trade Press, London, 1973.
16. Shapton, D.A., Lovelock, D.W. and Laurita-Longo, R., 'The evaluation of sterilization processes from temperature measurements in degrees Celsius', *J. Appl. Bact.* **34** (1971), 491.
17. Stumbo, C.R., *Thermobacteriology of Food Processing* (2nd edn), Academic Press, London, 1973.
18. Jowett, R. (ed.), *Hygienic Design and Operation of Food Plant*, Ellis-Horwood, Chichester, 1980.
19. *Food and Drink Manufacture – Good Manufacturing Practice: A Guide to its Responsible Management* (2nd edn), Institute of Food Science and Technology, London, 1980.
20. Paine, F.A. and Paine, H.Y., *A Handbook of Food Packaging*, Leonard Hill, Blackie, London, 1983.
21. Moody, B., *Packaging in Glass*, Hutchinson Benham, London, 1977.
22. Anderson, K.G. and Whiteman, W.E., in *Food Industries Manual* (2nd edn), Blackie, Glasgow, 1988, Chap. 17.
23. Dennis, C. and Rose, S., 'The microbiology of mayonnaise-based salads', *Proc. Inst. Food Sci. Tech. (UK)* **17** (1984), 69–78.
24. Troller, J.A. and Christian, J.H.B., *Water Activity and Food*, Academic Press, London, 1978.
25. Dakin, J.C. and Crawford, C., 'The efficacy of sorbic acid as a preservative for acetic acid products', BFMIRA Technical Circular No. 316, 1966.

8 Packaging

JOHN BETTISON

8.1 Introduction

Of the wide range of packages containing vegetables and vegetable products seen on the shelves of supermarkets and food stores today, the glass container has the longest history. Pressed glass vessels, capable of holding liquids and which are more than 3500 years old, have been found in Egypt, although the technique of glass-blowing – the logical technique for the production of containers suitable for the storage of foods and liquids – only appears to have been introduced some 2000 years ago, probably in Syria where the glass produced at that time was particularly suited to that purpose.

Mouth-blowing the glass into moulds in order to achieve consistency of shape and volume followed soon after (first century A.D.), and the manufacture of bottles and jars continued in this time-honoured fashion until late into the nineteenth century, when semi-automatic and, finally, automatic glass-blowing machines were developed.

Glass containers have been used commercially for the preservation of food products by heat for almost 200 years, and metal cans for the same purpose for almost the same length of time. The other packaging materials dealt with in this chapter – from paperboard to plastics to laminates of all descriptions – have been developed for food use within the last century, many indeed within the past 15 years; the rate of package development is constantly accelerating, but the more critical environmental climate leads to different approaches being made towards the re-use or disposal of the package itself.

Today, the packaging manufacturers regard themselves as partners with the food manufacturers in the food industry as a whole, and standards of manufacture and hygiene in package-making plants are constantly being uprated to equal those of their customers' food preparation areas. Regular inspections of such premises are made by food technologists representing the food manufacturers or major retailers, and one feature of BS5750: 1987 accreditation/certification is the ability of authorised representatives to visit a package-making plant, unannounced, at any time up to four occasions per annum, in order to re-establish conformity with the provisions of the BS 5750.

BS 5750 is the UK Standard for *Quality Systems and the Management of Quality Systems*, although it is as much a concept as a standard, in that it often requires a complete change in attitude and approach to quality. It is not about quality standards as such, that is, it does not lay down dimensional tolerances (such items have to be agreed by manufacturer and customer), but it does lay down fully documented procedures for actions to be taken by relevant personnel and designates one senior management representative for each factory/site to be responsible for ensuring that all such actions and procedures as stipulated are carried out. For example, a procedure drawn up for personnel involved in buying could include a statement that no goods will be purchased from any supplier which has not obtained, or applied for, formal BS 5750 accreditation.

Possession of formal BS 5750 accreditation is, in effect, a guarantee that the factory in question is operating a quality system which should ensure that agreed quality standards are maintained, and the major UK container and component manufacturers have now obtained such certification. Equivalent ISO International Standards (ISO 9000–9004: 1987) do exist, and are gradually being adopted by similar companies outside the UK.

8.2 Objectives of packaging and market considerations

8.2.1 *Packaging requirements*

The principal objective of food packaging is to hold the food product – ideally in the condition in which it was prepared by the food manufacturer – until it is ready to be consumed.

In order to achieve this ambition, it requires a combination of properties and/or characteristics, and although each recognised packaging material possesses some of these, sadly no one material possesses them all. The principal desired properties are listed below, although this cannot be considered an exhaustive list:

(i) It must conserve the contents within, free from microbial spoilage and be safe to eat.
(ii) It must maintain its integrity, that is, protect the contents from external attack by microorganisms and insect infestation, and be resistant to normal handling abuse.
(iii) It must protect the contents from sabotage or unauthorised opening or, if subjected to such abuse, it must be able to indicate that fact.
(iv) It should keep the food product in optimum condition, free from the pick-up of off-odours, off-flavours or taints and the nutritional value of the product must be retained within the shelf-life range demanded of it.
(v) Ideally, the packaging material should be impermeable to liquids and gases and not be attacked by the food product within.
(vi) It should not be expensive in relation to the value of the product contained therein.

(vii) It should be able to display the product, either through the walls or lid of the container itself or by illustration, and leave sufficient space for nutritional information on labels.
(viii) It should be easy to open, if opening instructions are followed, and not present a hazard with regard to its ultimate disposal.
(ix) It should be recyclable or recoverable/re-usable in some form.

Packaging does, however, have many more values than these, among which the most important may be considered to be the following:

8.2.2 *Economic benefits*

It has been estimated that 50% of all the food produced in the world in any one year is wasted; it deteriorates before it can reach a consumer. The processing and packaging of foods allows them to be preserved – usually at the point of production – so that they may be transported to distant markets where they may be sold in as good a condition as when they left the factory. The trend towards urbanisation, that is, the move away from 'the land', has accelerated this trend, and specialisation of crop growing adjacent to canneries and freezing plant in order to deliver the produce as fresh as possible, makes economic sense as well as resulting in a quality product.

8.2.3 *Marketing benefits*

Packaging allows for the extension of the shelf-life of the product within; this extension could be for a few days only in the case of fresh foods, but in others it could be many months, allowing the consumer to have access to food products 'out of season', bringing greater interest and enthusiasm to eating and cooking. In the context of marketing, shelf-life is an important consideration, as it is also in the context of labelling and legislation. The 'rainy day' storage mentality has now disappeared from the housewife's mind, and it is estimated that 95% of the canned foods packed in the UK are consumed within three months of their packing date – even though their shelf-life is well in excess of this time-span.

Another feature of packaged foods is the advantage which the marketer can take of package design or labelling to sell his product and project his company's image. Container shapes have particular appeal in the glass industry, and some products are instantly associated with a particular shape and colour of bottle or jar. The trend towards more convenient opening systems, for example, caps which screw off rather than need to be prised off, Ring-Pull easy open ends, peelable lids on plastics trays, is another way in which the packaging industry is trying to meet the demands placed upon it by its customers. But one must remember that the easier the package is to be opened, the easier it may be to abuse it, and then some tamperproof or tamper-evident feature needs to be built in.

One final feature on marketing is the need to develop, and then to publicise, systems designed to recycle/re-use packaging materials in the best interests of the nation as a whole.

8.2.4 *Long-term benefits*

Following the massive increases in energy and raw material costs – particularly for the plastics materials based on oil – of some 15 years ago, packaging manufacturers, like the whole of manufacturing industry, had to rethink their long-term strategy with regard to material supply. We have seen the introduction of thinner, double reduced, steel for cans, lightweight glass containers, the incorporation of reclaim into non-food contact areas of plastics containers and a trend towards recycling of all glass and metallic container materials.

Apart from such financial constraints, increasing environmental concern for our disappearing world, where everything comes out of the ground and some of it finds its way back there again as refuse; where litter is becoming a menace on a world scale – in the sea as well as on land – and where acid rain and waste gases pouring out of chimneys become reasons for marches and protests, has already led to major developments in packaging material manufacture and use, and such pressures will continue this trend. Reference will be made to specific items such as water-based lacquers, and ultraviolet curing of printing inks, but it is important that the whole of the food industry is aware of the initiatives being taken by the packaging industry to reduce waste, to maintain high quality standards and to eliminate the misuse and improper disposal of its wares after they have achieved their objective.

A whole range of packaging materials and container types is used for the marketing of vegetables, and these will be considered in the context of the type of process used to prepare the end-product, that is, whether it be heat-preserved, frozen, fresh/chilled or dehydrated. It has been decided to add the pickled products to the heat-preserved products, because the types of container used are essentially the same. Reference will also be made to aseptic foods in the same section.

8.3 Containers for heat-preserved vegetables

This has been a significant market sector for packaged vegetable products for many years and, perhaps for that reason, it has by far the widest range of packaging materials aiming for a part of this market. There are metal cans, glass jars/bottles, semi-rigid plastics containers (trays, tubs, pots and bottles), foil laminate pouches, aluminium foil trays, collapsible tubes and round transparent plastics 'cans'. These will be dealt with in turn, explaining the manufacturing process, methods of closing the filled containers and, where necessary, special considerations required for heat processing and subsequent handling of the containers. A final note will refer to containers for the aseptic packaging of vegetables and vegetable products, both in bulk and in retail-size containers.

8.3.1 *Metal cans*

The publication, in Paris in 1809, of Nicholas Appert's work *The Art of Preserving*, led to the development of what we now know as 'the canning industry', even though Appert's experiments were undertaken using glass jars sealed with corks. A copy of this work, translated into English, formed the basis of British Patent No. 3372, dated 25 August 1810, issued to Peter Durand; Durand, however, in his specification, extended the range of containers for which the heat-preservation technique was suitable, to include 'vessels of glass, pottery, tin or other metals or fit materials'. In 1811, Durand sold his patent to the firm of Donkin, Hall and Gamble, who decided that glass was too fragile and they then began to use soldered tinplate containers, which they commenced manufacturing commercially in Bermondsey, London, probably in 1812. At that time, a good tinsmith was able to manufacture only about ten cans per day from sheets of tinned iron.

8.3.1.1 *Tinplate.* Moving to more recent history, until the 1930s tinplate was manufactured by dipping steel sheets into baths of molten tin – hot-dipped plate. Later in that decade, however, a continuous tincoating process was developed with the plate from a coil feeding through electroplating baths – electrolytic plate. This technique gave better dispersion of the tin, so that lower overall coating weights could be used with equivalent protection, and it also permitted the deposition of different amounts of tin on either side of the steel coil – differential electrolytic plate; the vast majority of tinplate now used for vegetable can manufacture is differential plate, with a tincoating of 5.6 grams per square metre (gsm) frequently being used for the outside of the can to minimise rusting/external corrosion problems and a tincoating of 2.8 gsm frequently being used for the inside of the can (but see Section 8.3.1.4).

Apart from the gradual reduction in the amount of tin used on vegetable cans, there have been moves to reduce the amount of steel used in the cans also in order to contain cost increases. Within limits, it is possible to reduce the thickness of the steel and maintain its strength – as long as the temper or hardness of the steel is simultaneously increased; this is normally achieved by subjecting the steel to an additional cold-rolling process – double-reduced plate. Unfortunately, the increase in temper resulting from cold rolling is very directional, which imposes some restrictions on can-making techniques capable of handling such plate. Mechanical methods of strengthening the bodies of metal cans exist such as the incorporation of circumferential beading; this increases their panelling resistance but reduces their axial load resistance, so that compromise settings must be achieved following extensive field trials between the packer and canmaker.

8.3.1.2 *Tin-free steel.* If tinplate may be regarded as 99% steel and 1% tin, tin-free steel (TFS) could be regarded as 99.9% steel and 0.1% chromium. Thus in all physical and mechanical respects, it behaves in a similar manner to tinplate, and it is essentially only in its surface chemical performance that it differs. TFS is produced by the electrodeposition of chromium and chromic oxide only a few molecules thick on to the steel surface, giving a material which is able to perform satisfactorily in some packaging applications at marginally lower cost than an equivalent tinplate specification. It is now very widely used for the manufacture of caps and closures and the ends seamed on to tinplate or TFS-bodied food cans.

8.3.1.3 *Aluminium.* In comparison to the aforementioned steel-based materials, aluminium is about one-third the weight and a much more ductile material. An alloy of pure aluminium with manganese and magnesium is required to impart strength to the material before containers and components suitable for use with vegetables and vegetable products can be prepared.

When preparing aluminium sheet for food can/component applications, it is important that care is taken throughout all stages of its manufacture, from ingot preparation, control of rolling speeds and lubricants, in order to ensure that extremely high surface quality standards are maintained.

In general terms, food containers and components made from aluminium are not as resistant to mechanical abuse such as handling and stacking as steel cans, and aluminium also loses some of its inherent strength when subjected to high temperatures such as may be encountered during the oven curing of applied lacquers or external coatings and decoration. Therefore, it may be necessary to redesign canning line layouts, outer packing for distribution and warehousing facilities when considering the introduction or changeover to aluminium containers or components from steel.

8.3.1.4 *Lacquers.* Although cans were manufactured for many years without lacquers, that is, they were internally plain, consumer demands and legislative pressures – as well as economic considerations – have led to the almost universal adoption of fully internally lacquered steel and aluminium components in metal food cans. With the tinplate container, the use of a lacquer will commonly permit a significant reduction in the weight of the tincoating required, whereas with TFS, lacquers are necessary – both internally and externally – in order to reduce toolwear during the fabrication of components therefrom.

Apart from these advantages, internal lacquers can add to the consumers' quality perception of the food itself by the use of a gold or white internal coating, and materials such as zinc oxide or zinc carbonate may be sparingly incorporated therein in order to absorb volatile sulphur compounds from proteinaceous foods such as vegetables and thereby reduce the possibility of the production of visually unacceptable sulphide staining within such cans –

zinc sulphide is opalescent whereas iron sulphide is black. Modern lacquers used for cans and components designed for vegetable products are usually based on epoxy-phenolic or PVC resins and give excellent performance.

Lacquers for most two-piece draw redraw (DRD) can bodies, for three-piece can bodies and for ends to be used with vegetable cans of whatever construction, are applied to the flat sheets of metal as a solution/dispersion in an organic solvent on high-speed coating/lacquering units linked to gas-heated travelling ovens which may be up to 50 m in length. During passage through these ovens, which may be at temperatures over 200 °C for several minutes, the solvents are driven off (to be treated in special incinerators before being harmlessly released to the atmosphere) and the lacquer resins are polymerised to form a 'bond' with the metal surface. In the interests of economy and reducing the possibility of atmospheric pollution, there is growth in the use of the so-called water-based lacquers – with lower levels of volatile organic components – and the use of ultraviolet radiation to cure lacquers, with the consequent elimination of the large curing ovens (see Figure 8.1).

An alternative to lacquering, namely the lamination of plastics films to either or both sides of tinplate, TFS or aluminium is now becoming available, with potential uses for DRD cans and component manufacture.

8.3.1.5 *Can manufacture.* There are three common methods of making cans suitable for the processing of vegetables, namely:

- the three-piece can with a welded side-seam body;
- the two-piece DRD can;
- the two-piece draw wall-ironed (DWI) can.

The former technique, which is still the most widely used food can-making technique, at present is applicable to tinplate only on an economic basis, whereas the DRD technique is suitable for tinplate, TFS or aluminium; the DWI technique is already being used with tinplate and aluminium, for example, for beverage cans, but at present only in tinplate for food cans primarily because of body strength requirements.

Three-piece welded can. The three-piece electrically welded food can may be manufactured on can-making lines which were originally designed to make soldered cans, with little modification other than the replacement of the bodymaker units (see Figure 8.2). After stencil-lacquering of the sheets of tinplate, they are fed into a slitter, which cuts them into pieces, or body blanks, identical in size to the final can body cylinder. The bodymaker rolls these blanks into cylinders and the two edges of this cylinder are drawn together so that they butt with minimum overlap. At this stage, the edges are welded together electrically using continuous copper electrodes, and a coat of lacquer (internal side-stripe) is applied internally by roller or spray over the welded

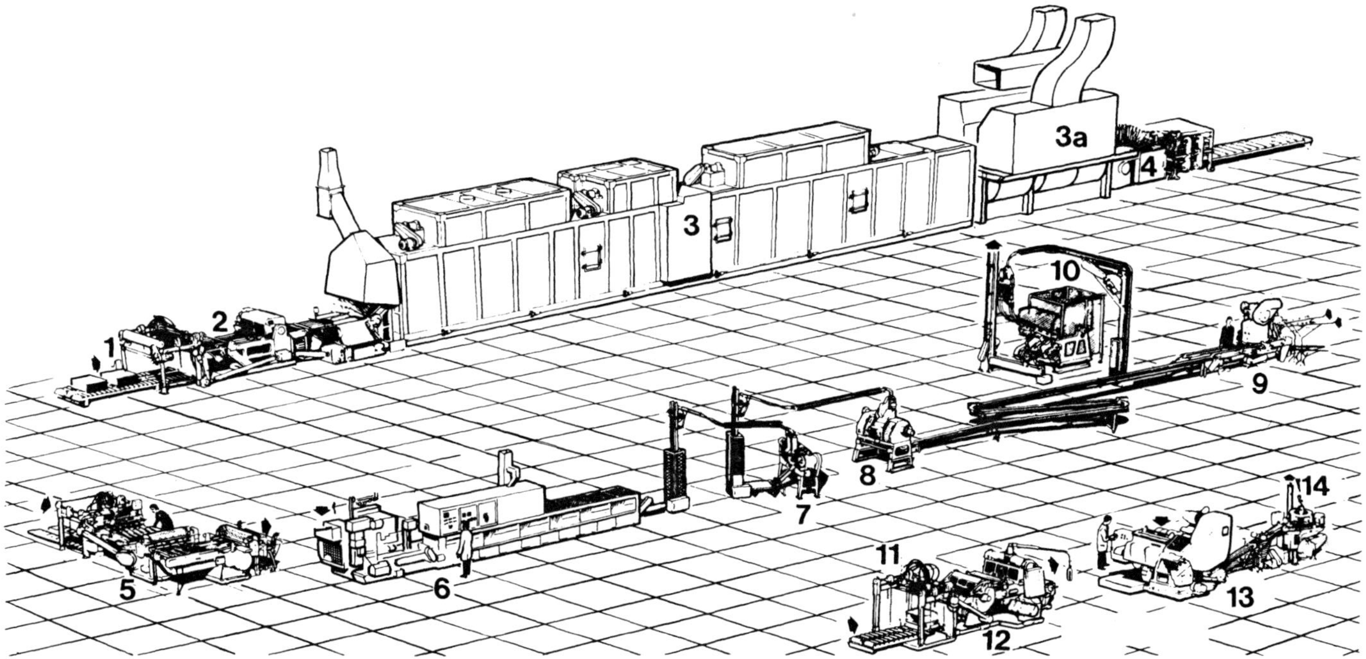

Figure 8.1 Sheet coating line (1–4): *1* Sheet feeder; *2* Lacquering/coating unit; *3* Curing oven; *3a* Cooling section; *4* Sheet stacker.

Figure 8.2 Can making line (5–10): *5* Slitter; *6* Bodymaker/welder; 7 Can divider; *8* Flanger; *9* Seamer; *10* Pressure tester.

Figure 8.3 End making line (11–14): *11* Sheet feeder; *12* Scroll sheer; *13* End press; *14* Compound liner. (courtesy CMB Packaging).

seam and then cured in an oven. The body cylinders are beaded, where necessary, flanged and transferred to a double seamer where one end of the can (known as the maker's end) is double seamed on to the cylinder (see Figure 8.4). Double seaming of welded cans is easier than it was with soldered cans because on soldered cans, at the overlap, there were two thicknesses of plate plus the thickness of the solder between, whereas with the welded can, the side-seam has a thickness of less than 1.5 times the plate thickness. Double seaming is a critical operation and requires regular control during production.

End manufacture again consists of cutting strips of plate from the lacquered sheets by a scroll-shear device, stamping out and forming ends from the lacquered circles therein and lining them with a solvent-based or water-based lining compound in order to effect a hermetic seal on the can body cylinder by double seaming (see Figure 8.3). The manufacture of full-aperture easy-open ends is more sophisticated and requires closer control of plate specification and lacquering in order to achieve consistent performance. However, for

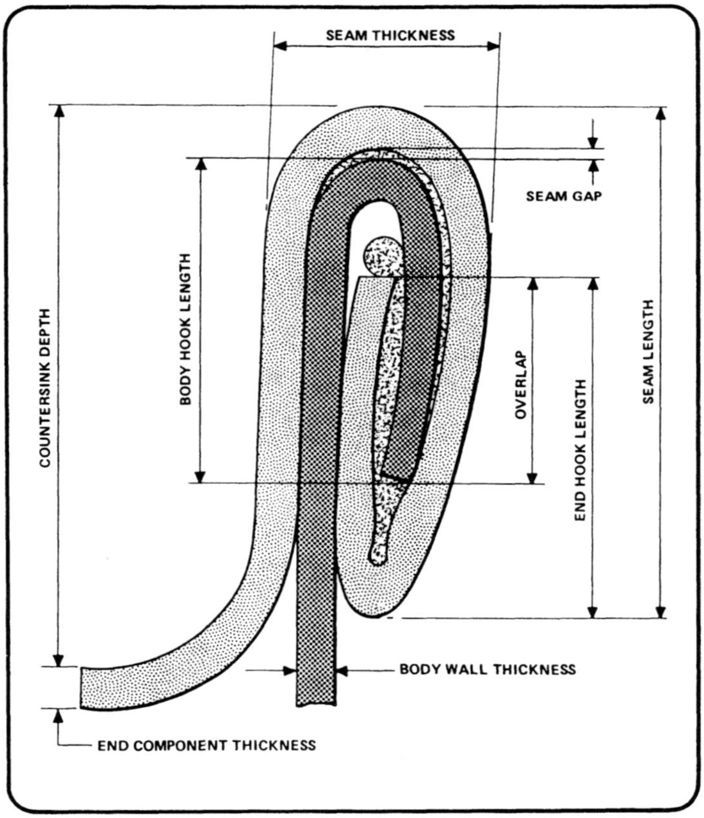

Figure 8.4 Double-seam dimensional terminology (courtesy CMB Packaging).

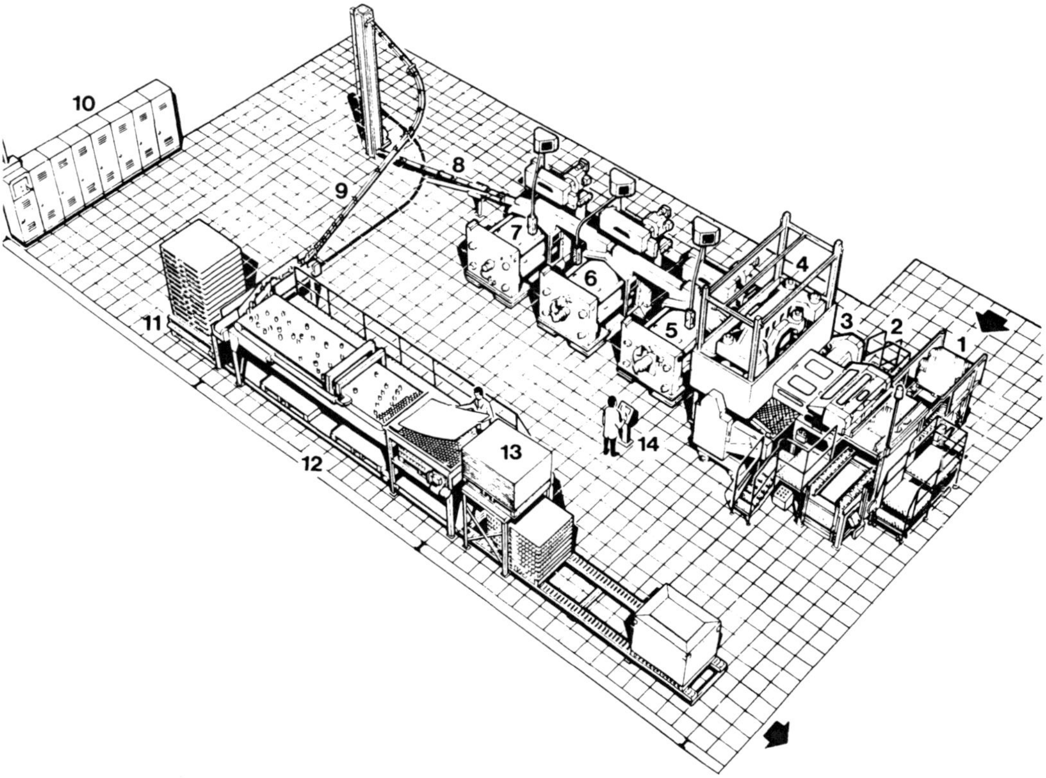

Figure 8.5 Two-piece DRD can line: *1* Infeed stillage conveyor; *2* Lubricating station; *3* Sheet feeder; *4* Blank and cupping press; *5* 1st redraw rotary press; *6* 2nd redraw rotary press; *7* Trimmer rotary press; *8* Can sampling chute; *9* Elevator and chute; *10* Line control cabinets; *11* Pallet dispenser; *12* Can packer; *13* Cardboard packing; *14* Line packer console. (courtesy CMB Packaging)

vegetable products, there are no technical reasons why they should not be more widely used.

Two-piece DRD can (see Figure 8.5). The DRD technique has been used for general container manufacture for many years, but not until quite recently for the manufacture of cans which are taller than their diameter. With this system, the body and bottom end of the can are made from a circular disc of metal, lacquered/coated on both sides, the area of the disc being approximately the same as the surface area of the final container into which it is to be made, that is, the thickness of the base and wall of the container are identical and equal to the original thickness of the disc itself.

The disc undergoes a number of cupping/drawing operations which reduce the diameter and increase the height and which finally result in a trimmed,

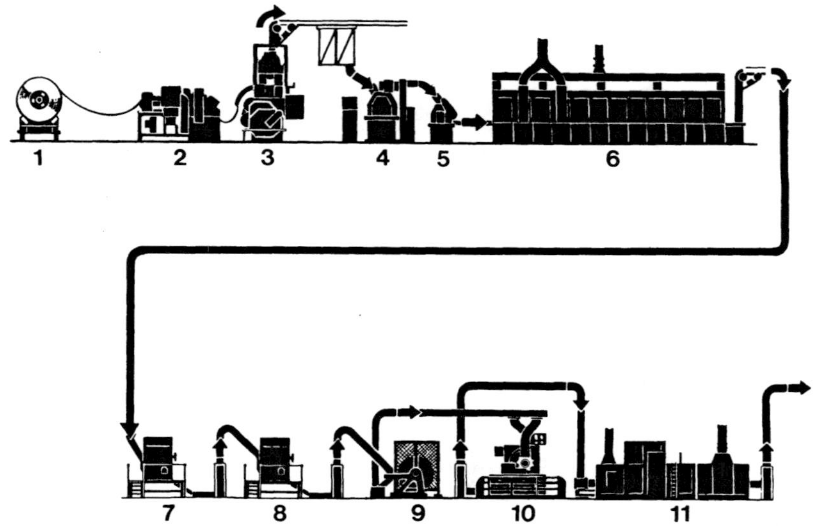

Figure 8.6 Two-piece DWI can line: *1* Coil unwind; *2* Inspection; *3* Cupping press; *4* Bodymaker; *5* Trimmer; *6* Washer; *7* Flanger; *8* Beader; *9* Tester; *10* Spray lacquering; *11* Curing oven. (courtesy CMB Packaging)

flanged can complete with bottom end incorporating expansion rings. DRD cans are normally very strong containers, capable of resisting high internal vacuum levels due to their thick body walls, and for this reason are widely used for products which are normally closed under a very high vacuum regime, for example, sweet corn. DRD cans may easily be made in a variety of shapes – round, square, rectangular, oval – although for vegetable canning applications, the round can is still the favourite.

Two-piece DWI can (see Figure 8.6). The DWI technique is almost universally used for the manufacture of cans for beer and soft drinks, where the internal pressure developed by the carbon dioxide in the product itself helps to support the filled can during its storage and distribution. With this technique, a shallow cup is made from a disc of thick plate, and the walls of the can are then ironed (stretched and thinned) through concentric metal rings until the final height of the can being made is achieved. The can body is beaded, trimmed and flanged, and is then ready for despatch to the packer. The finished can has a base equal in thickness to the original metal, incorporating expansion rings, whereas the walls are typically less than one-half that thickness.

The degree of working necessary on the base metal during ironing demands the presence of sophisticated lubricants, and these need to be washed off prior to final internal spray-lacquering operations. Pre-lacquering of the metal disc is not feasible with the DWI technique.

Because of the relatively weak body wall which results from the DWI manufacturing operation, the use of such cans for foods tends to be limited to relatively solid products which may be successfully packed at low levels of vacuum, in order to avoid can body wall collapse, or to products which are able to afford and support the adoption of a nitrogen/liquid nitrogen injection technique to provide internal pressure for the container.

One advantage of the two-piece can, whether DRD or DWI, is the fact that there is only one seam on the can (the canner's end), rather than three seams on the traditional three-piece can (the maker's end, the side-seam and the canner's end) and that there is no difference in bodyplate thickness around the flange of the two-piece can whereas the side-seam – even on a welded three-piece can – is significantly thicker than the remainder of the can body.

8.3.1.6 *Corrosion of metal cans.* Fortunately, vegetables cannot be considered as one of the most corrosive of food products packed into metal containers. Reference has been made earlier to the potential problem of sulphide staining which may arise if care in the selection of an internal lacquer system is not taken, and problems with tomato products, especially tomato paste, which may cause hydrogen swells in inappropriate can specifications, are not unknown. Otherwise, modern can-making and lacquering techniques have effectively solved internal corrosion problems, and external problems such as rusting can be controlled by paying attention to correct retort venting, can drying and water treatment.

8.3.1.7 *The future for metal cans.* People have been forecasting the demise of the metal can for many years, but the present health and vigour of the market suggest that it will be with us for at least the commodity vegetable items for many years to come. The continued development of new methods of reducing the raw material inputs into can manufacture will ensure that it has an economic future at least, and Easy Open end developments will increase the convenience of the package to the consumer. They are being used with all forms of heat-preservation techniques, and their inherent strength and reputation for reliability will continue to serve them well. On the question of recyclability, it is perhaps of interest to note that, in 1988, 55% of all aluminium beverage cans made and filled in the USA were recycled.

The introduction of plastics-laminated steel/aluminium materials, eliminating the need for extensive lacquering/coating/oven curing facilities, may write a whole new chapter in the history of can-making and usage of cans for foods.

8.3.2 *Glass containers*

Glass containers have been used commercially for the heat preservation of food products for almost two centuries. Indeed, the first processed food containers which were used by the Frenchman, Nicholas Appert, late in the eighteenth century were made of glass. Since those early days, when natural corks were used to seal the glass jars, the closure industry had introduced a variety of caps and closures for glass containers.

8.3.2.1 *Properties of glass.* As a container for heat-preserved vegetables and vegetable products, glass is perhaps the perfect choice; it is inert, impervious to gases, odours and flavour components, and resistant to chemical attack – certainly by chemicals of vegetable origin. Glass containers may be transparent, allowing the consumer the opportunity to examine the product before purchase (although for light-sensitive products, green or amber glass may be used).

Although glass in its newly formed state is twice as strong as steel, unfortunately in the form in which it is normally found in real life its strength is only one-thousandth of that original value. Fragility must therefore be considered as a disadvantage possessed by conventional glass containers used for the packaging of vegetable products.

Glass is a poor conductor of heat, and sudden changes of temperature in excess of 60–65 °C in container glass may cause potentially dangerous stresses due to thermal shock; however, sudden heating is less dangerous than sudden cooling in this respect. This factor does demand more care during the heat processing of glass food containers when compared to conventional metal cans, although not necessarily so compared to the newer plastics-based processable containers.

It is important to note, however, in this context that the glass container is eminently suitable for use in microwave ovens, once any metal closure has been removed, and this adds to its image of convenience.

One final disadvantage for glass containers is their weight compared to alternative packaging materials, but developments are under way to reduce this; a recently introduced 'lightweight' glass jar was some 30% lighter than the existing jar, yet gave similar performance with regard to handling and abuse resistance.

8.3.2.2 *Manufacture of glass containers* (see Figure 8.7). Glass containers for use with heat preserved vegetables and vegetable products are normally made from clear, or flint, glass. If required, green or amber glass may be produced, or even opal glass for some traditional products where a 'stoneware' image needs to be maintained.

Ingredients. The three principal raw materials used in glass container manufacture are sand (SiO_2), soda ash (Na_2CO_3) and limestone ($CaCO_3$). Silica sand is the main ingredient, but because of its very high melting temperature, soda ash and limestone are added to give a formulation which melts at a furnace temperature around 1500 °C. Traces of other ingredients such as felspar, calumite and selenium may be added to the mixture depending

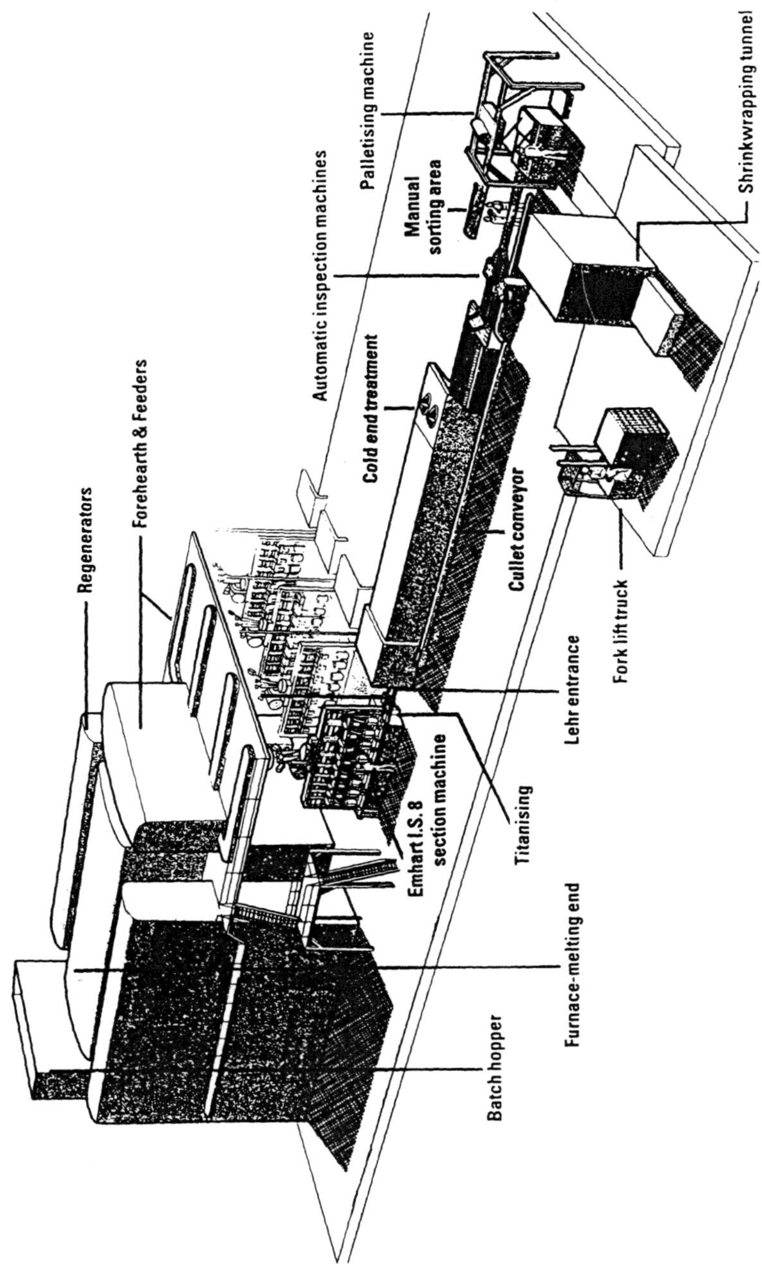

Figure 8.7 A typical furnace (courtesy United Glass Ltd).

on the type of impurities present in the three bulk raw materials. (For example, where a high iron content is present, selenium must be added, otherwise a green glass will result.)

A typical mixture for glass containers might include 60% sand, 18% soda ash, 17% limestone and 5% mineral additives.

Finally, a proportion of cullet (carefuly sorted and washed broken glass) is mixed with the other materials. The incorporation of cullet has a number of advantages, the main one being economic, in that it reduces the energy required to form glass in the furnace from the other raw materials. It is estimated that the incorporation of 10% cullet into a batch reduces the energy costs in glass container manufacture by 2%, and it is now common amongst European glass-makers for 80% of the total mixture to be cullet in batches of green container glass, and 30% in batches to be used for clear glass container manufacture. This cullet arises either within the glass-making plant itself as scrap or reject bottles and jars, or as recycled glass collected from Bottle Banks.

This availability of recycled glass and the fact that the other raw materials are among the commenest constituents of the earth's crust stands glass in good stead for the future, always bearing in mind its high demand for energy.

Batch plant. The purpose of the batch plant, now computer-controlled, is to calculate the required level of ingredients/cullet for the type of glass being made and to feed it slowly by a helical screw feed system to the furnace.

Furnace. Almost all container glass is now melted by the continuous process in regenerative furnaces, in which the ingredients slowly enter at one end and molten glass is slowly withdrawn from the other end. The heat is provided by burners using oil or natural gas, and the heat regeneration system, which recovers heat from the waste gases and uses it to pre-heat the combustion air, conserves energy.

The operating temperature within the furnace is around 1500 °C, and is maintained day and night. After five to six years of such treatment, the furnaces are normally demolished and rebuilt. Because of the loss of heat through the furnace walls, furnaces are becoming ever larger, and some are now able to deliver in excess of 500 tonnes of molten glass per day to the container-forming machines.

Forehearth and feeder. The molten glass is conveyed from the furnace to the forming machines via a forehearth and feeder (or gob-forming unit). The forehearth is a covered channel made of refractory brick, which is fitted with heaters and cooling units so that the glass can be presented to the feeders at a temperature controlled precisely to within 1 °C. The feeder forms the glass into gobs (beads of molten glass of even weight and uniform temperature around 1100 °C) by shears which cut it to an accurate length as it emerges through an

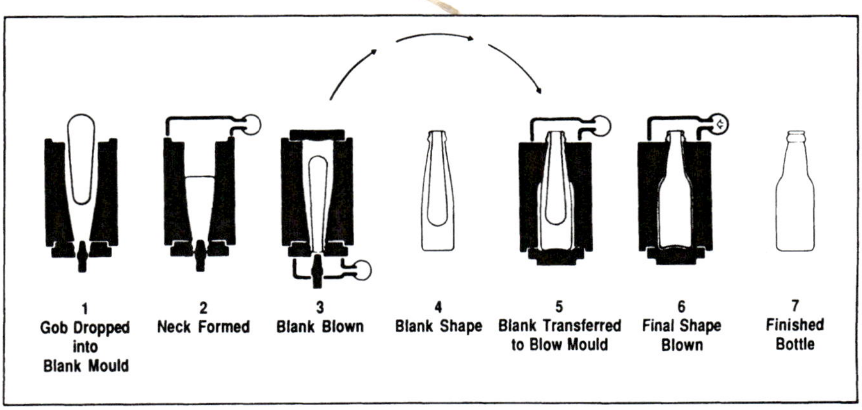

Figure 8.8 Glass container forming: Blow and blow – for narrow-mouth containers (courtesy United Glass Ltd).

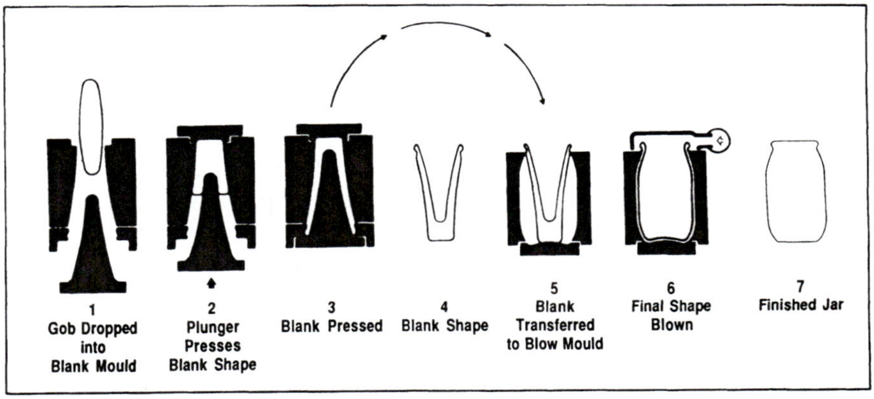

Figure 8.9 Glass container forming: Press and blow – for wide-mouth containers (courtesy United Glass Ltd).

orifice, and distributes these via chutes to the forming machines. The size of the gob is governed by the size and weight of container being made and under conditions of high line efficiency, can be maintained within a weight range of plus or minus 1%.

Container-forming machines (see Figure 8.8 and 8.9). The IS (Independent Section) machine was introduced as early as 1925 for the forming of glass containers, and is still used today. Modern versions of this machine can now produce containers at speeds in excess of 300 per minute and handle over 100 tonnes of glass per day. Either one, two or three containers can be made on each section, and the machine itself can comprise up to ten sections.

The gob of glass is loaded into a 'parison' mould which forms the 'blank' shape of the container, and is then transferred to a blow mould before being blown out to its final shape. Within the framework of the IS machine are two processes for making containers, namely 'blow-and-blow' and 'press-and-blow' although, because of the greater degree of dimensional control achievable with the latter technique, this is now being used almost exclusively for vegetable and vegetable product jars and bottles.

The diagrams show the differences in the two methods, but with the press-and-blow technique, the gob of glass, after dropping into the parison mould, is forced by a plunger which forms the body cavity and simultaneously presses the glass into the required neck finish. This partially formed container is then transferred – by its neck – to the bottle mould, where the final shape is obtained by air pressure applied through the neck of the bottle.

Annealing lehr. The containers leave the forming machine at a temperature of approximately 650 °C and, if allowed to cool rapidly, would develop internal stresses and be more susceptible to impact damage. For this reason, they are rapidly conveyed to a temperature-controlled tunnel or 'lehr', through which they slowly pass and in which they are slowly reheated and then cooled at a predetermined rate.

Surface treatment. The trend towards the lightweighting of glass containers has necessitated the increasing use of glass surface treatments in order to maintain the overall 'strength' of the containers by permitting their smooth flow through packing lines and improving their abrasion resistance.

The first stage, or 'hot end' treatment, is applied on the conveyor between the forming machines and the annealing lehr; it normally consists of spraying the hot bottles or jars with a vapour of organic titanium or inorganic tin compounds and is considered as being able to double the strength/resistance of the container.

The second stage, or 'cold end' treatment, is applied to the containers at the exit of the annealing lehr and normally consists of an organic compound such as oleic acid, which increases the lubricity of the containers and enables them to move efficiently through high-speed filling lines.

Glass neck ring finishes (see Figure 8.10). For every type of closure used on glass containers, there is an appropriate glass neck ring finish. The glass-makers and closure manufacturers work very closely together in this area and, once the type of closure is decided upon and the design of the container (height, diameter and shape) chosen, the glass-maker will fit the appropriate diameter neck ring finish demanded by the closure.

8.3.2.3 *Closures for glass food containers.* Closures for glass food containers have, in general, a number of similarities. Normally, they are made of

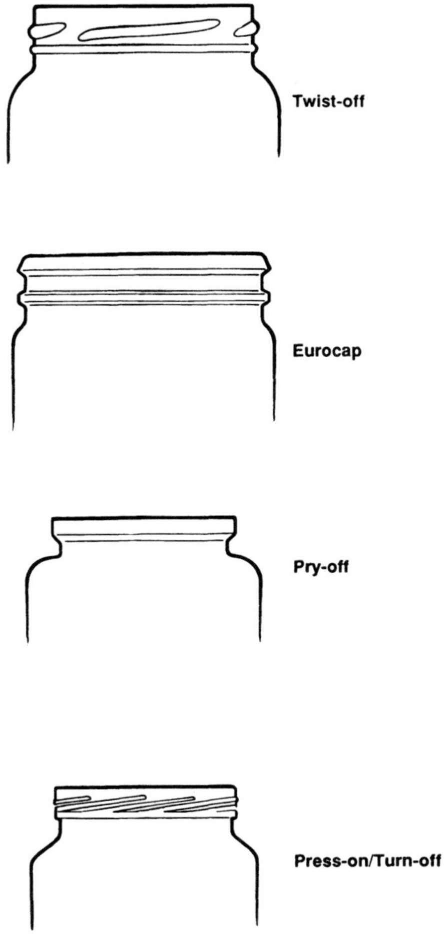

Figure 8.10 Glass neck ring finishes.

aluminium, tinplate or tin-free steel (TFS), with the venting closures primarily being made of aluminium and the non-venting closures of tinplate/TFS; they are lacquered or coated both internally and externally to protect the metal from corrosion or attack by the product itself, and to enhance the sales appeal of the package. Finally, they are usually provided internally with a ring of lining compound which in some closures also extends down the skirt of the closure.

Closure manufacture. The manufacture of metal closures basically consists of the following operations:

(i) The coils of metal are cut into sheets of convenient size for handling through the printing and lacquering units.

(ii) These sheets are passed a number of times through printing presses and coating units which apply inks and varnishes to the outside of the sheet and product-resistant lacquers to its inner surface. The choice of lacquer(s) used depends on the type of food product to be packed. Between each of these operations, the sheets are passed through a heated oven, which is able to cure the coated materials and bake them on to the surface of the plate, often at temperatures in excess of 215 °C.

The above operations are performed on the flat sheets of metal prior to any forming or shaping of the closure itself.

(iii) The printed, lacquered sheets are then slit into strips ready for feeding into stamping presses, which cut them into discs, press and curl them into shape; liquid lining compound, which creates the sealing gasket, is then injected into the shell of the cap while rapidly rotating. This rotary action serves to flow the compound into the channel of the cap, although with some type of closure it is also necessary to mould the compound to encourage it to arrive at the appropriate position; a final heat treatment is given in an oven in order to cure and set the compound, often at a temperature in excess of 230 °C.

Closure manufacturers, like the glass manufacturers, have set themselves up to achieve BS 5750 certification for their operations.

Closure classification. Closures for glass containers for foods may be divided into two broad categories, namely the venting closures and the non-venting closures.

In the simplest of terms, a venting closure may be defined as a closure which is simply crimped on to the jar and which then permits, by controlled venting, the escape of air entrapped in the jar during subsequent heat treatment. With the onset of the cooling phase, a vacuum is created in the jar, this vacuum pulling the closure down and thereby sealing it on the neck of the jar.

By contrast, a non-venting closure is applied to the jar under vacuum, thereby giving an immediate hermetic seal, and the closure is maintained in this fully sealed condition throughout subsequent heat treatments by the application of a controlled overpressure to more than balance the internal pressure developed by the product within the jar itself.

Venting closures. As regards venting closures, typical of this range may be mentioned the Omnia cap and the Pano range of closures. These closures, which are normally made of aluminium, are not hermetically sealed on the jars when applied, the application of the cap being achieved by clinching the rim and/or skirt of the cap around the neck of the jar. Under normal closing machine setting conditions, these closures will start to vent when the internal pressure in the jar reaches 0.2 kg/cm^2. Critical machine setting is essential in order to give acceptable cap venting conditions.

Non-venting closures. As regards non-venting closures, typical of this group may be mentioned the Whitecap style of closure. These closures, which are normally made of tinplate or TFS, are hermetically sealed on to the jars, the application of the cap normally being achieved under vacuum by the injection of steam into the headspace of the jars and surrounding the closure with steam during application. The vacuum created in the jar headspace is sufficient to prevent the venting of the closure during normal pasteurising conditions at less than 100 °C, although a calculated overpressure is necessary during sterilisation of such jars at temperatures in excess of 100 °C. Calculations of the appropriate overpressure may simply be made by reference to a pressure diagram, from which the maximum internal pressure generated in the jars is established by knowledge of the product filling temperature, the level of vacuum created during closing, the headspace in the jars and the process temperature being used.

Of the non-venting closures, the Twist Off style closure, commonly available in diameters between 30 and 110 mm, is the most popular; it is applied to multithreaded glass finishes. The PT (Press on/Turn off) closure differs from the Twist Off closure in that it is applied simply by pressing it down on to the jar without any rotary action, since it is not fitted with lugs; because of the placement of the lining compound down the skirt of the closure itself and the fact that this compound takes up the shape of the threads on the glass neck ring during pasteurisation or sterilisation, it is possible to screw the closure off. Whitecap style closures, when used for sterilised products, are not normally subjected to temperatures above 125 °C; a maximum total retort pressure of 2.45 kg/cm^2 should not be exceeded.

Regarding the comparison of venting and non-venting closures, venting closures generally require less sophisticated glass neck ring finishes and may perform better than non-venting closures with inconsistently or badly filled jars, whereas non-venting closures are hermetically sealed on the capping machine and remain in that condition for the remainder of their lives; such closures may also be applied at significantly higher speeds and in general possess better opening characteristics.

Both types of closure are widely used on pickled/pasteurised vegetables and vegetable products packed in glass within Europe, among examples of which are the large quantities of onions, gherkins and other products packed both in eastern and western Europe. For the relatively more sophisticated sterilised products such as asparagus, carrots, and garden peas, non-venting closures are normally used, whilst the non-venting Deep Twist Off or Medium Twist Off closures in 30 and 38 mm diameters are widely used on glass bottles of tomato ketchup and other sauces. For the pasteurised and sterilised vegetable products referred to above, closure diameters of 58, 63, 66 and 70 mm are commonly used for the 0.5 kilo capacity jars, whereas 82 mm is normally adopted for the larger 1 kilo jars.

Some slight inroads into the pickled vegetable market will be made by plastics closures – perhaps on PET or similar plastics jars – but this is not yet seen as a serious threat to the quality end of the market, where the major future growth is expected. It is possible, however, that combination caps with a metal top sealing disc and a plastics tamper-evident skirt will find a niche in this market.

Handling of glass vegetable containers. In some ways, glass containers are more difficult to handle than conventional metal cans, especially at high speed, but as long as filling lines are designed to take this into account and care is exercised throughout all operations, there is no reason for concern in a well-run plant. There are five areas in the plant and distribution system which are most important in ensuring that the jars reach the consumer in perfect condition, namely:

(i) *Filling conditions.* Clean filling, without contamination of the jar sealing surface, is necessary, as is a consistent filling temperature and, for non-venting closures, a consistent headspace volume of not less than 6% of the jar brimful capacity.

(ii) *Capping conditions.* Capping machines used for non-venting closures are available in a range of operating speeds from 10 to 1500 containers per minute. The caps are orientated in the correct condition to seat above the jar neck by a rotating hopper equipped with pick-up pins or magnets; the caps are then fed through a steam-heated chute to the point in the capper where they are picked up by the jar. Steam is injected into the jar headspace and the cap is finally pressed or rotated on to the jar neck ring finish by overhead sealing belts.

It is necessary to control the performance of the capper, and this is done by checking the in-jar vacuum levels on cold-water filled jars, by checking the vacuum on product-filled jars and by performing, on lug-style Twist Off caps, a security check to measure the tightness of application.

Venting style closures are normally applied on rotary crimp-type machines, not under vacuum, and because they do not necessarily incorporate a reclosure feature, the aforementioned checks do not apply.

(iii) *General jar handling.* It is particularly important with jars to avoid cap-to-cap contact, which could disturb the hermetic seal between jar and cap, and the glass containers themselves should be designed to eliminate the possibility of such contact during passage of the containers along filling lines and between side guiderails. Conveyor speeds and deadplates should be set up so that containers do not make heavy contact with each other, and for vacuum-closed jars the use of in-line dud detectors will remove unsaleable and damaged jars from production lines.

(iv) *Processing conditions.* Care has to be taken to ensure that no unacceptable stresses are placed on the closure during heat processing. Different procedures are required to prevent this depending upon the types of heat treatment being given to non-venting style closures as follows:

(a) *Hot filling.* This is used most commonly with vegetable sauce and juice products such as tomato ketchup and tomato juice. The only potential problem here is thermal shock failure of the glass were it to be filled at 85–90 °C while itself being at a temperature of, say, 20 °C. It is normal practice to pre-heat the containers to 50–55 °C by means of a rinse with hot water at that temperature.

(b) *Pasteurisation.* Pickled vegetables, such as onions and gherkins are normally filled at ambient temperatures and pasteurised at temperatures up to 75–85 °C. With such products, it is important to achieve as high an internal vacuum level at the time of cap application as possible, so that any internal pressure built up with the increase in product temperature resulting from the pasteurisation treatment is minimal – most of the non-venting closures will accept some internal pressure before they leak.

For venting closures, this cold-filling technique facilitates the production of a good internal vacuum on cooling.

(c) *Sterilisation.* During heat processing at temperatures in excess of 100 °C, the actual internal pressure built up in a glass container closed with a non-venting closure is governed by a combination of filling and capping conditions and the sterilising temperature being used. Such jars are normally sterilised under water with superimposed air pressure, although other overpressure systems such as steam/air (e.g. Lagarde), water/air (e.g. Steriflow) and overpressure hydrostatic units (e.g. Stork, Hunister) are equally suitable. Pressure calculator charts exist which allow the calculation of the internal pressure built up within the jars, these calculations being based upon four factors which are measured prior to the commencement of processing, namely:

- the internal vacuum in product-filled containers;
- the product filling temperature;
- the percentage headspace volume in the filled containers;
- the processing temperature to be used.

Having established the internal pressure, it is necessary to operate the retorts with a total pressure within them 0.3 kg/cm^2 higher than this; there are limitations to the amount of pressure which certain types of closure can withstand without damage, and if these limits are exceeded in the calculations above, it will be necessary to alter one or more of the control parameters, for example, increase the product fill temperature, in order to achieve successful production.

The retort overpressure conditions pertaining to the use of venting closures follow an essentially similar pattern to the aforementioned.

(v) *Warehousing and storage.* The final outer packing operation and subsequent warehouse storage of filled stock is also important in maintaining the processed containers in optimum condition. Excessive top loading must be avoided in order to minimise disturbance of the caps and the lining compound within. Points to watch with regard to this are as follows:

(a) Where the jars are being shrinkwrapped for distribution, it has been found that microflute board trays give better resistance to abuse caused by excessive top loading.
(b) The pallets used must be in good condition and palletised loads must be lowered gently in position. It is also recommended that wooden or chipboard divider pads be placed between each stacked pallet if single-faced pallets are being used.
(c) Ideally, pallets should not be stacked more than three high in warehouses for long-term storage, although four-high stacking for shorter periods is satisfactory under carefully controlled conditions.

8.3.3 *The retort pouch*

The retort pouch was developed in the 1960s in the USA by a consortium of food packaging/processing companies working on a contract from the US Army and in conjunction with the US Army Natick Laboratories. The pack has become most widely accepted in Japan, although it finds application in Europe, particularly for vegetables and vegetable-based recipe products.

Advantages claimed for the retort pouch include its thin cross-sectional area which allows more rapid heat transfer than a cylindrical metal or glass container of equivalent volume. It is estimated that with the retort pouch, the total process time may be reduced to about 40% of that for an identical product of the same volume in a cylindrical can. Other advantages include:

- Low weight and volume (both empty and filled) resulting in savings of both storage space and transport costs.
- High opening convenience: the pouch may be torn open or simply cut with scissors.
- Boil-in-the-bag reheating: as many pouches as necessary may be reheated together in one saucepan; packs may be lifted out of hot water by hand.
- Ease of disposal: packs may simply be crushed for disposal.

Packing materials. In construction, the retort pouch was originally supplied pre-formed with three of the heatseals ready made by the pouch-maker. But most modern lines receive pouch material in reel form for making into pouches on form-fill-seal equipment by the packer himself. The material used for the manufacture of the retort pouch is normally a three-ply lamination consisting of a tough outer ply to resist snagging or abrasion, a middle layer to provide a barrier to oxygen, light and moisture, and an inner ply to provide good heatsealing and product compatibility. A typical example of such a material would have a polyester outer ply, an aluminium foil barrier and an inner ply of polypropylene, having the facility to be sterilised at temperatures in excess of 121 °C if required.

Filling and processing. Although foil laminate pouches have been filled and sealed on high-speed automatic equipment for some years, handling products of the type packed into retort pouches (for example, hot vegetables, two-stage filling of solids and liquids) presented new problems, which were finally overcome by the adaptation of form-fill-seal principles. Because of its structure, the retort pouch cannot be processed in steam or under water without an overriding air pressure. The pressure of saturated steam at normal processing temperatures is not enough to balance the internal pressures developed inside the pouches – even when conforming to the recommended maximum headspace of less than 2% of pouch volume. The resulting differential pressures could give rise to 'ballooning' and possible bursting of the pouch resulting from a weakening of the seals. Ballooning also adversely affects the rate of heat transfer into the pouches and the efficient distribution of the heat transfer medium within the processing chamber.

8.3.3.1 *Post-process handling of pouches.* Unlike the can, where post-process infection is unlikely when the can is dry, the pouch could possibly become infected even when in that condition. Any small 'pinhole' in the laminate could result in the transfer of product to the surface when the laminate is 'flexed', which could occur if the pouches are handled. Contamination from surface, aerial or human contact could be effected and spoilage result. For this reason, the following Code of Practice was drawn up and recommended by Campden Food and Drink Research Association and the pouch suppliers for the post-process handling of retort pouches:

(i) The use of chlorinated water for cooling the pouches.
(ii) The use of mechanical post-process handling equipment which must be carefully sanitised; processed pouches must be mechanically handled at all stages up to and including cartoning/overwrapping.
(iii) Retort pouches must be dried as quickly as possible after cooling.
(iv) Retort pouches must be individually overwrapped or cartoned.

At its launch, the retort pouch was seen as creating an entirely new market for food products not at that time packed in cans, a market more akin to frozen foods than existing processed foods. Unfortunately, the package concept itself

(a) INJECTON MOULDING

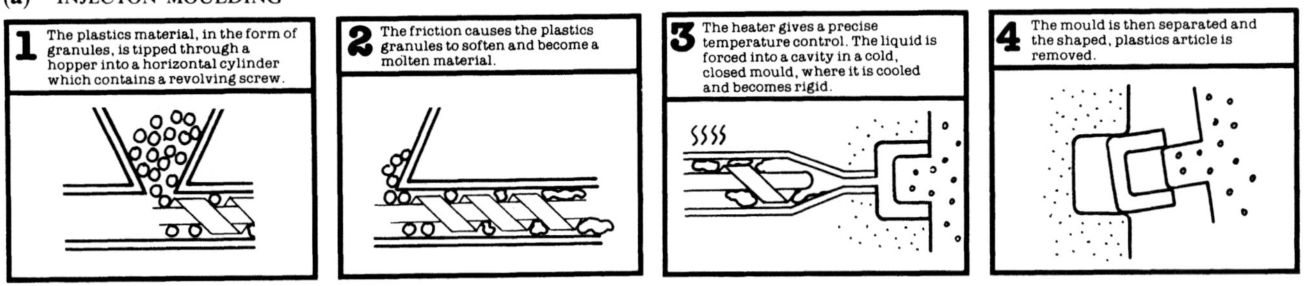

(b) BLOW MOULDING

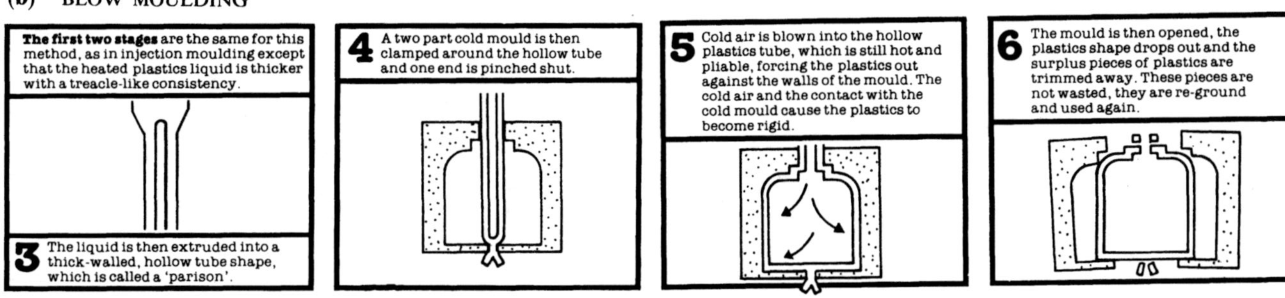

(c) THERMOFORMING

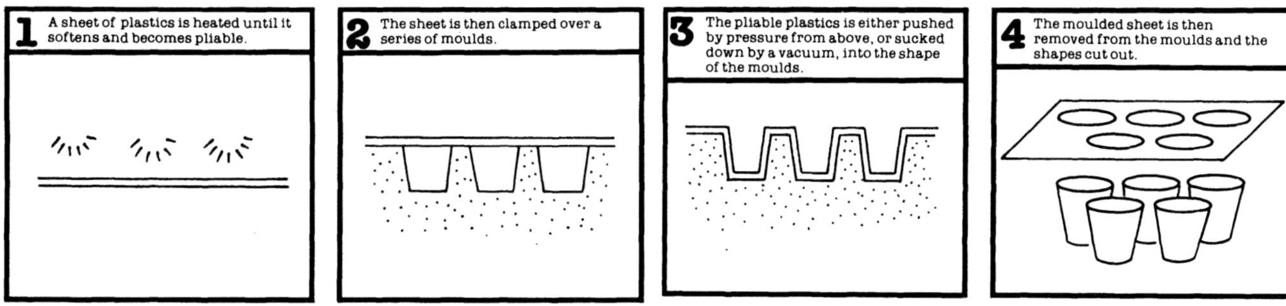

Figure 8.11 Polypropylene. (a) Injection moulding. This method is used to make plastics containers, closures, tubes and dessert goblets; (b) Blow moulding. This method is used mainly for bottles and some wide-necked jars; (c) Thermoforming (pressure forming). This method is a simple process used for certain basic shapes such as tubs, pots and lids, i.e. the simpler and more shallow shapes which do not need to be particularly strong. (courtesy CMB Packaging).

did not appeal to the mass of the public and its market impact has not been strong.

8.3.4 *Rigid/semi-rigid plastics containers*

The use of plastics containers for the packaging of heat-preserved foods has increased significantly during the past ten years, albeit from a very small base. It may perhaps be considered as the most significant packaging development of the twentieth century because the possible replacement of the traditional metal can and the glass jar – the accepted 'processed' food containers for almost 200 years – by plastics containers must be of fundamental worldwide interest. Although some of these containers are designed to run through existing can- or glass-filling lines, however, the incontainer processing technique may have to be modified from that used with conventional metal cans, and the closing technique used may be different from that used with the traditional containers.

8.3.4.1 *Plastics materials.* In order to give adequate performance as a container for foods which are to be sterilised, or even pasteurised, after filling, the packaging material from which the container is made must possess both adequate heat resistance (that is, normally 121 °C for sterilised foods and 85–95 °C for pasteurised foods) and adequate barrier performance, particularly against moisture and oxygen transmission.

Unfortunately, no single plastics material yet offers the ideal of low price, easy forming characteristics, heat resistance and barrier performance, and so materials need to be combined, normally by co-extrusion, in order to approach this ideal. Among the wide range of plastics materials now available, the four which at present find widest application in this packaging area are:

(i) *Polypropylene.* Polypropylene has a high softening point (140–150 °C) and is therefore suitable for sterilising operations – even at temperatures above 120 °C. Its resistance to the passage of oxygen is poor. It can be blow moulded and injection moulded, although it is relatively difficult to thermoform (see Figure 8.11).

(ii) *PET.* Polyethylene terephthalate (PET) can be blow moulded into high clarity bottles and tubes. Its gas barrier properties are good, but its heat resistance is relatively poor unless a heat setting technique is used; modifications to that technique will allow containers incorporating PET components to be used for hot filling, pasteurisation or even sterilisation.

(iii) *Barrier materials – PVDC and EVOH.* The gas barrier performance of a plastics container may be improved by increasing its wall thickness (not usually an economic or structurally sound approach), by coating it with PVDC (polyvinylidene chloride) or by incorporating PVDC or EVOH (ethylene vinyl alcohol copolymer) within a co-extrusion. Other barrier materials are under development, but the aforementioned are the most important at the present time. For coating application, an external coating applied by dipping, roller coating or spraying is the preferred system, using a water dispersion of PVDC.

The choice between PVDC and EVOH as a barrier layer in a coextrusion or laminate is not easy, but EVOH at present appears to be favourite on the basis of recyclability, that is, incorporation, as regrind, into new containers, of scrap from the container forming operation. Methods of improving the recyclability of PVDC, however, are constantly being sought and some barrier plastics are being used for this purpose.

8.3.4.2 *Container manufacturing techniques.* Basically, there are four types of plastics container being used or pack-tested for heat-preserved vegetables and vegetable products, namely:

(i) Sterilisable barrier plastics containers.
(ii) Hot-filled barrier plastics bottles.
(iii) PET-bodied metal ended cans.
(iv) Monolayer plastics jars/tubs.

Sterilisable barrier plastics containers. In order to achieve the barrier performance at economic cost, a thin layer of PVDC or EVOH is extruded as one layer of a coextruded web formed via a co-extruding die. The multilayer sheet thus formed may consist of polypropylene inner and outer layers, the barrier layer next to the inner layer and a reclaim layer between the barrier and the outer layer. The reclaim layer consists of the reground skeletal web which remains from the original multilayer sheet following the removal of the finished containers therefrom (see Figure 8.12).

In the manufacture of Lamipac® containers, the containers themselves are thermoformed from the sheet material; other forming techniques may be used for the production of alternative style containers. Lamipac® containers, after filling, may either be closed with metal ends using a sophisticated double-seaming technique, or a foil/plastics laminate (for example, PET/foil/polypropylene) may be heatsealed on to the containers using the TOR process, which removes headspace air from the pack by a vacuum and pressure technique prior to heatsealing (see Figure 8.13).

This removal of air assists the processability of the pack (in individual layers, under water or steam/air processing procedures are ideal at temperatures up to 121 °C) and extends product shelf-life. For vegetables and vegetable products, a shelf-life of 18–24 months is achievable in such containers in temperate climates, the vegetables being blanched before filling.

Fused or peelable seals may be produced, as desired. A typical rate of oxygen uptake for a 450 g Lamipac® container, after processing, would be 1.0 ml per month, but it is again important to bear in mind that the TOR process results in essentially a nil oxygen headspace situation at time of closing. One convenience feature of some such containers, as well as their

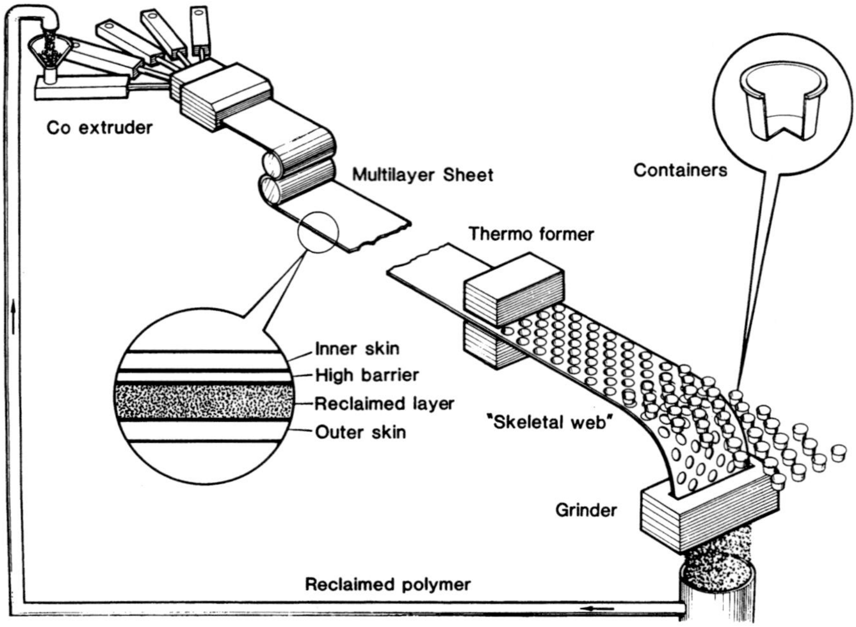

Figure 8.12 Lamipac container manufacture (courtesy CMB Packaging).

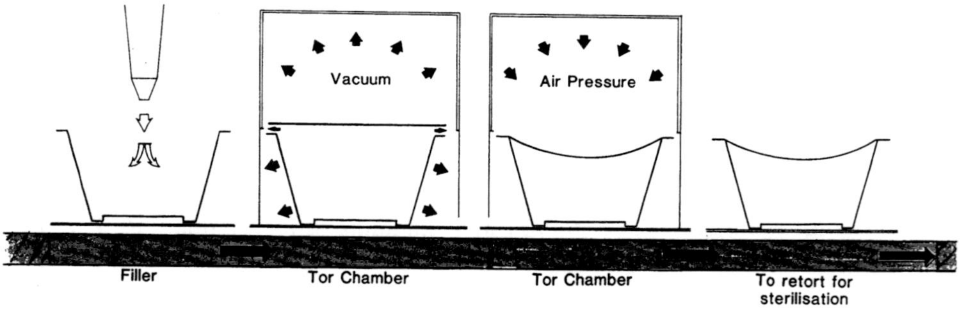

Figure 8.13 Heat sealing by the Tor process. Air removal for long shelf life creates a hydraulically solid pack for ease of heat processing (courtesy CMB Packaging).

peelable seals, is the fact that they are dual-ovenable, being reheatable in both microwave and hot air ovens. Decorated outer cartons or overwraps are normally used with these containers.

As well as Lamipac® containers, a wide range of sterilisable multilayer plastics containers now exist, either commercially or under trial, among which may be mentioned Ultralock, Altradec, Menu-Box, Omnican and Lunch Bucket.

Hot-filled barrier plastics bottles. As with the aforementioned materials, the required barrier performance is again achieved by the incorporation of a thin layer of, for example, EVOH, but this time using the co-extrusion blow moulding technique, whereby a multilayer parison is produced by the use of two or more extruders coupled to one die.

A typical example of such a hot-fillable bottle is the Lamicon bottle, which is now widely used for the packaging of tomato ketchup and similar products.

Such containers use a five- or six-layer structure (see Figure 8.14) and for hot fill applications, polypropylene is used as the inner and outer ply, with adhesive tie layers anchoring the EVOH between them. If reclaim is to be incorporated within the polypropylene layers, a further layer of virgin polypropylene may be required as the food contact layer for some applications. After hot filling, the necks of the bottles are closed with a heatseal-coated aluminium foil disc beneath a plastics (dispensing) closure. Heatsealing may be performed through the closure itself by induction heatsealing techniques.

The advantages offered by such plastics containers for tomato ketchup and similar products are unbreakability of the bottles and convenience of product dispensing, in that just the required amount may be squeezed out of the bottle. A shelf-life of 18 months for hot-filled ketchup is regarded as achievable, while a wide-mouth 2.5 litre Lamicon jar of similar construction, heatsealed with coated foil and overcapped, is also capable of giving similar shelf-life performance to sweet pickled vegetables. A similar style container to the Lamicon bottle is the Gamma bottle.

PET-bodied metal ended can. Although it has been found that even biaxially oriented PET containers will tend to shrink or distort when exposed to relatively low temperatures (even as low as 65 °C), it has also been established that if such a container is exposed to elevated temperatures under controlled conditions, without permitting it to shrink, it will become 'heat set'. Such heat-set PET containers may then be exposed to conventional in-container sterilising temperatures without significant shrinkage or distortion occurring. A typical example of such a container is the Stepcan (see Figure 8.15).

The Stepcan basically consists of a PET body cylinder, made by the Stretch Tube Extrusion process, to which two metal ends are double seamed, one after the product is filled. PET granules are extruded to form a tube, the tube passes through a conditioning process and it is then biaxially oriented (stretch/blown). Heat setting is performed by means of a patented hot mould/cold mandrel technique in order to impart the required process temperature resistance into the finished container. The resulting can body cylinder is glass-clear and possesses gas barrier properties sufficient to give a shelf-life of 12 months with certain products. A longer shelf-life may be achieved by resorting to PVDC coating techniques.

Metal can ends are seamed on by a modified double-seaming operation, and

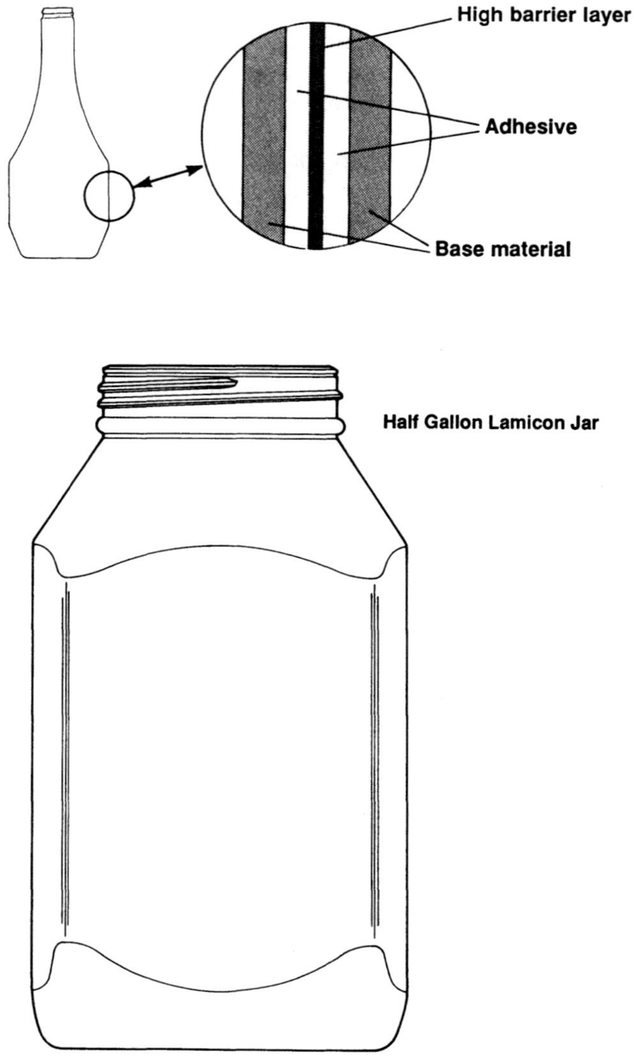

Figure 8.14 Multilayer structure of Lamicon container (courtesy CMB Packaging).

cans may be run through conventional canning lines with modifications, although a significantly different approach to filling procedures is required in order to avoid bulging or panelling of the can bodies after heat treatment.

Stepcans have been used commercially with a range of pasteurised acidified vegetable products, and trials with sterilised vegetables are being evaluated. The use of a glass-clear body cylinder on the Stepcan essentially combines the best features of both of the traditional processed food containers, namely the

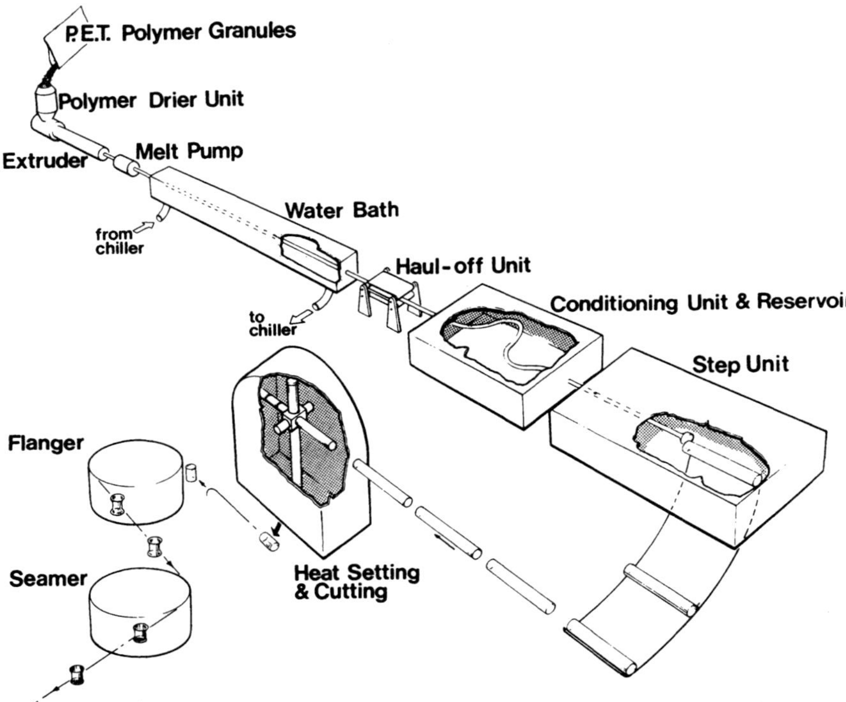

Figure 8.15 The Stepcan® (courtesy of CMB Packaging).

transparency of the glass jar with the potential high speed line handling characteristics and 'unbreakability' of the metal can.

Monolayer plastics jars/tubs. There is a trend towards the introduction of transparent PET jars to replace glass and translucent polypropylene containers to replace cans for certain products. The shelf-life performance of such containers is limited, as is their resistance to temperature (in the case of PET), but it is possible that a niche exists in certain markets for such containers with a limited range of products.

8.3.5 *Aluminium foil trays*

Such semi-rigid trays are normally manufactured from aluminium foil 0.1–0.14 mm thick bearing a polypropylene coating 0.015 mm thick, by a simple pressure-forming technique. They have been used as sterilisable containers for a limited range of food products including meat and vegetables. The trays are filled and hermetically closed by heatsealing an 0.05 mm coated foil lid across the top; the trays may then be sterilised at 121 °C in retorts under water, or in steam/air, in overpressure conditions. The containers are very susceptible to damage or pressure distortion and for that reason, may be packed into printed outer cartons to improve their sales appeal. Such containers are claimed to be dual-ovenable, that is, reheatable in both conventional hot-air ovens and in microwave ovens when the lid has been removed.

8.3.6 *Collapsible tubes*

The use of collapsible tubes for foods, although not as yet common in the UK, is widely accepted for a limited range of products on mainland Europe and beyond. Such tubes are normally manufactured from aluminium, although co-extruded high barrier plastics laminates are being considered. Typical vegetable products which are packed in this way are tomato paste and mustard, the products being in a shelf stable condition under ambient or chill refrigerator conditions.

Aluminium tubes are manufactured by an extrusion process, wherein a 'slug' or disc of aluminium is forced through a die along principles not unlike the manufacture of the DWI can referred to earlier, except that the disc used for the tubes has a hole in its centre which is formed into the screw thread cap sealing area. The open-ended tube resulting therefrom is trimmed to the desired length, softened by annealing and internally lacquered. A screw cap or closing device is fitted to the thread sealing area, and the tube is then ready to be filled through its base.

After clean filling with product, normally pre-heated to achieve commercial sterility and cooled, the open end of the tube is folded over and crimped/heat-sealed to produce a hermetic seal. Subsequent storage/distribution conditions depend on the type of product packed.

Heat sterilisable versions of aluminium collapsible tubes have now been developed, but their use has still not extended much beyond the world of astronauts.

8.3.7 *Aseptic packaging*

Aseptic packaging may be defined as the filling of a commercially sterile product into a pre-sterilised container in a sterile atmosphere, and applying a pre-sterilised cover or seal to the container – also in a sterile atmosphere – in order to obtain a hermetic seal.

Aseptic packaging has been around for many years and is now very widely used for a limited range of products, for example, dairy products, sweets and puddings, and fruit juices, and less commonly but increasingly used for vegetable products such as tomato paste, tomato and other vegetable juices, tomato sauce bases, and soups. Methods of preparing products aseptically have already been discussed, and the increasing ability to fill particulate products is significant.

8.3.7.1 *History.* The process of aseptic canning was invented by Dr W.M. Martin and developed commercially by the Dole Engineering Co. of California, USA, some 40 years ago. In the original equipment, conventional three-piece tinplate cans were pre-sterilised by superheated steam; filling and double seaming of the cans was carried out under aseptic conditions in a steam-filled chamber – the can ends having received a similar dry heat sterilisation to the empty cans. In the UK, the aseptic canning technique has been used commercially for a limited range of products such as custard and ice-cream mix, and reference is made to it here mainly for historic reasons, because it never appeared to offer many advantages in the vegetable packaging market.

There are two significant market areas, however, where the aseptic packaging of vegetable products does offer advantages both with regard to higher quality and potentially reduced costs, namely the use of retail-size packs for vegetable soups, vegetable juices and tomato sauce bases; and the use of bulk aseptic packages for the trans-shipment of tomato paste from one country to another.

8.3.7.2 *Retail-size containers.* Of these, the most widely used containers for vegetable products are Tetra Brik and Combibloc cartons, usually in capacities of 250 ml, 0.5 litre and 1 litre. There are significant differences in the principles of each carton system, although to the final consumer these may not be all that apparent. The impact that aseptically filled cartons have had world wide in the 25 years since their introduction may be gauged by the forecast that some 35 000 million such packs were sold in 1989.

8.3.7.3 *Tetra Pak aseptic carton system* (see Figure 8.16). The Tetra Pak system, developed in Sweden, was introduced in aseptic tetrahedral form as early as 1961, with the rectangular Brik version following in 1969. The system uses reel-fed, pre-printed, pre-creased, laminated material, the laminate normally comprising seven layers (from inside to outside) as follows:

polythene – polythene – aluminium foil – polythene – paper – ink – polythene.

The material passes over a series of rollers which soften the web. It is then sterilised by passage through a bath of hot 30% hydrogen peroxide. It then enters at the top of a sterile chamber, and as it descends it is formed into a cylinder and sealed longitudinally. The product is introduced into the tube just prior to this point and the transverse seals are formed by a set of jaws which press the tube together (thereby squeezing product from the seal area) and then make the seal. The filled pack is cut off and it is finally formed into its rectangular Brik shape. The filling area, after being sterilised by hydrogen peroxide and heat, is maintained in a sterile condition by a continuous overpressure of sterile air (Class 100 air produced by HEPA (a high efficiency filter system) or cartridge filters in a laminar flow unit is normally satisfactory).

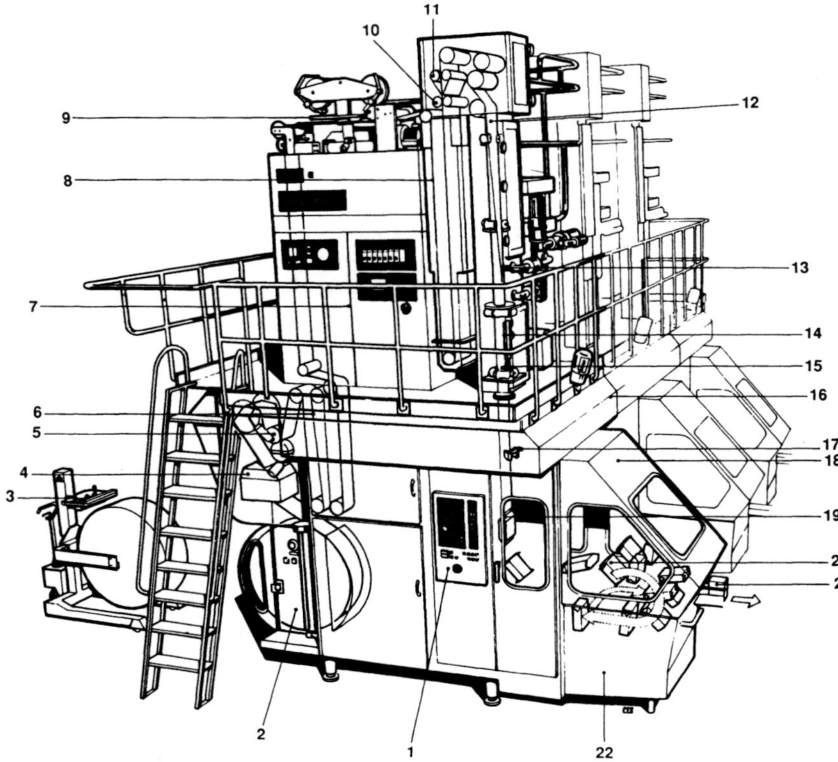

Figure 8.16 Tetra Brik aseptic packaging machine, TBA/8. *1* Control panel; *2* Roll with packaging material; *3* Trolley with hydraulic lift for transport of packaging material; *4* Automatic splicing equipment for packaging material; *5* Stamping unit for best before date; *6* Loops of packaging material which ensure smooth and jerk-free material feed and make possible continuous production when new packaging material is to be spliced on; *7* Most of the electronic control system of the machine is placed here; *8* The packaging material is sterilised in a deep bath of heated hydrogen peroxide; *9* Strip applicator which provides one edge of the packaging material with a plastic strip which, at the subsequent longitudinal seam sealing operation, is welded together with the other edge. This results in a tight and durable seal; *10* Squeezer rollers which remove the hydrogen peroxide from the packaging material; *11* Nozzles for hot air which dries the packaging material; *12* Here, the packaging material begins to be formed into a tube; *13* Filler pipe; *14* Longitudinal seam element which welds together both edges of the packaging material; *15* 'Short stop element' which completes the longitudinal seal when the machine is restarted after a brief production stop; *16* The machine is designed for interconnection to form compact production units where two or more packaging machines have a common platform; *17* Photocells which control the artwork retention; *18* Raisable and lowerable hood over the automatic washing system and the final folder section where the top and bottom flaps of the cartons are folded in and sealed against the cartons; *19* Sealing of the cartons takes place below the level of the liquid using induction heat. This is supplied by a pair of jaws which also shape and cut off the cartons; *20* In the final folder section, the top and bottom flaps of the cartons are sealed on two lines; *21* Discharge of finished cartons; *22* Batch which is automatically filled with water and detergent in conjunction with outer washing of the machine. (courtesy Tetra Pak Ltd)

Sealing 'through' the product has the advantage of leaving no headspace in the finished pack, for example, with tomato juice, but it does present a potential problem in that it could exclude the filling of fibrous or particulate products which could interfere with seal integrity. However, Tetra aseptic lines have successfully handled crushed whole tomatoes with a total soluble solids of 8–9% without apparent sealing problems.

8.3.7.4 *Combibloc aseptic carton system* (see Figure 8.17). The Combibloc system (PKL/Jagenberg) uses pre-made carton blanks, normally made from polythene/paperboard/polythene/aluminium foil/polythene, which permits

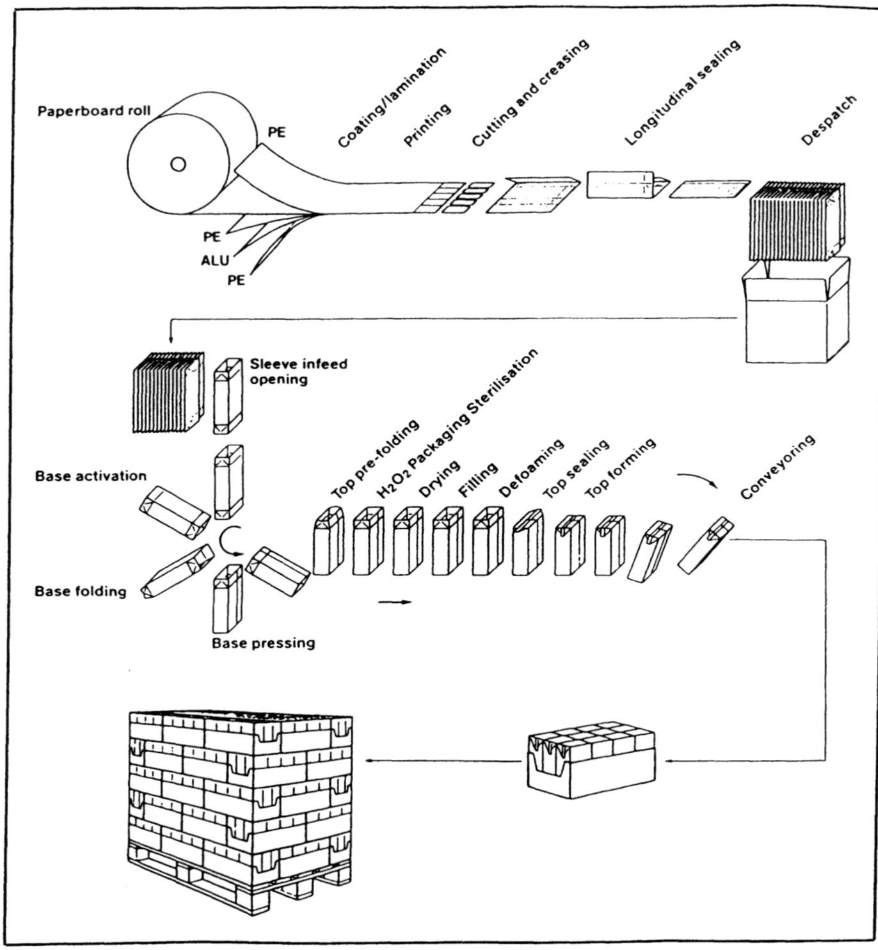

Figure 8.17 Combibloc (courtesy PKL).

the use of different size cartons on the same unit as long as they are of the same cross-sectional area. A carton blank is extracted from the magazine, its base is sealed, the top is pre-folded and it finally enters the aseptic zone where a positive pressure of sterile air is maintained. Hot hydrogen peroxide is sprayed into the carton, dried with hot air at 200 °C, and the carton is then filled, steam-flow closed and sealed across the top by ultrasonic welding. The presence of a small headspace in this carton is claimed to be an advantage in that it allows sealing even when particulate products are being filled, and it allows the product to be shaken before opening. The slight vacuum in the container reportedly enhances the shape and stability of the pack.

A number of Combibloc aseptic lines are now handling a tomato sauce base product which consists of whole crushed tomatoes with a total soluble solids of 8–9% without difficulties, and the range of vegetable soups now includes tomato with mushroom, pea soup and green bean soup.

The aforementioned carton packs will give a shelf-life of at least 12 months with vegetable products.

8.3.7.5 *Other retail-size aseptic packs.* There is also a very wide range of aseptic plastics container systems available (pots, tubs, trays), but basically

Figure 8.18 CMB Freshfill ML4 unit (courtesy CMB Packaging).

these may be grouped together into three different types, as follows:

(i) *Pre-formed monolayer containers.* These may be pre-printed, normally to a very high quality. Shelf-life is limited to the (poor) barrier performance of the material being used, for example, polypropylene or even polystyrene. Typical units built to fill this type of container are the CMB Freshfill SL1 and ML4 units (see Figure 8.18) and units from Gasti or Remy. After hydrogen peroxide sterilisation, the pots are aseptically filled and finally sealed with a pre-sterilised aluminium foil diaphragm.

(ii) *Pre-formed multilayer barrier containers.* These are handled on machines identical to (i) above, but offer a shelf-life of 12 months or more depending upon the barrier materials used.

(iii) *Form-fill-seal multilayer barrier containers.* These consist essentially of conventional form-fill-seal units with addition of hydrogen peroxide immersion baths for the base and top webs. The thermoformed cups are filled in a sterile tunnel and then lidded. Typical aseptic form-fill-seal units are the Erca, Formseal and Servac systems. Other systems are also available.

It is possible to add to these systems yet a further range of aseptic plastics bottle filling/closing systems, for example, of Remy, Sidel or Serac origin, which may be used for the aseptic packaging of vegetable juices.

8.3.7.6 *Bulk aseptic systems.* Bulk aseptic packaging systems are of growing importance in the worldwide distribution of products such as tomato paste, fruit purées and fruit juice concentrates. The increasing cost and 'inconvenience' (especially of opening and disposal) of using 5-kilo cans for transporting such commodity items over long distances has encouraged this trend, apart from the improvement in product quality (taste and colour) which result.

There are two principal bulk systems, one employing 're-usable' metal drums and the other relying on a plastics or laminate bag within a board (wooden or composite) outer. Such units, when of appropriate size, are normally provided with an International Standards Organisation pallet base to ease mechanical handling.

Drum systems. Such systems, of which the Cherry-Burrell Sterilpac and the Fran Rica systems are examples, normally use 55-gallon drums. The drums are pre-sterilised with wet steam, filled with cooled sterile product and then sealed under aseptic conditions. Typical filling speeds are up to 18 drums per hour.

Bag systems. The bulk bag-in-box systems, of which Scholle, Rossi e Catelli and again Fran Rica are examples, use gamma-radiation sterilised bags (up to 25 kGy) which, after aseptic filling in a sterile atmosphere, may be placed in a supporting box, carton or drum. Bag capacities range from 6 to over 1000 litres. Laminated bag materials, including foil laminates, may be used to give improved barrier properties and longer shelf-life, a common material now being a metallised polyester/polythene laminate.

Systems have been developed which permit the aseptic removal of the aseptic product from the bulk pack, thereby allowing its direct use in subsequent aseptic filling operations.

8.4 Packaging of frozen vegetables

Preservation of food by freezing is one of the oldest methods known to man, although the birth of the 'consumer' frozen food industry dates back to 1915, when Clarence Birdseye developed quick-freezing processes and equipment.

Prior to the use of a quick freezing process, a short heat treatment (blanching) at or below 100 °C is normally given, the main purpose of which is to inactivate enzymes. Enzyme activity is only retarded by normal quick-freezing storage temperatures, some enzymes still being active at temperatures as low as − 75 °C. Following the sophistication of the range of packaging materials used for heat-processed vegetables, it is rather simpler to cover the properties required for packages for frozen foods.

It may be convenient here to outline the principal method used for the manufacture of polythene film, and to show how film laminates are produced (see Figures 8.19 and 8.20) as these materials are referred to throughout the remaining sections of this chapter.

Frozen food packaging has to protect the product first and foremost from dehydration or drying out by providing an effective barrier to moisture vapour loss from the food surfaces during storage. Fortunately, most packaging materials are better moisture barriers at low temperatures than at ambient temperatures. Dehydration during freezing leads to the condition known as 'freezerburn' which gives an unsightly blotchy appearance to the product. Polythene is the best moisture barrier of the readily available packaging materials. A polythene bag of frozen peas, for example, might be expected to lose less than 0.01% moisture after three months' storage. Secondary effects of moisture loss include weight loss, with its potential for the creation of illegal weight declarations and a potential increase in the rate of oxidation of the product.

Protection from oxygen is an advantage, particularly for delicately flavoured products, although the rate of oxidation at the very low temperatures encountered during freezing and subsequent storage is sufficiently low to ensure that oxygen transmission is not a major problem to most products. It is of course essential that the packaging material used can withstand the low temperatures encountered during freezing and storage without deterioration. Of the plastics materials, polythene has high abuse

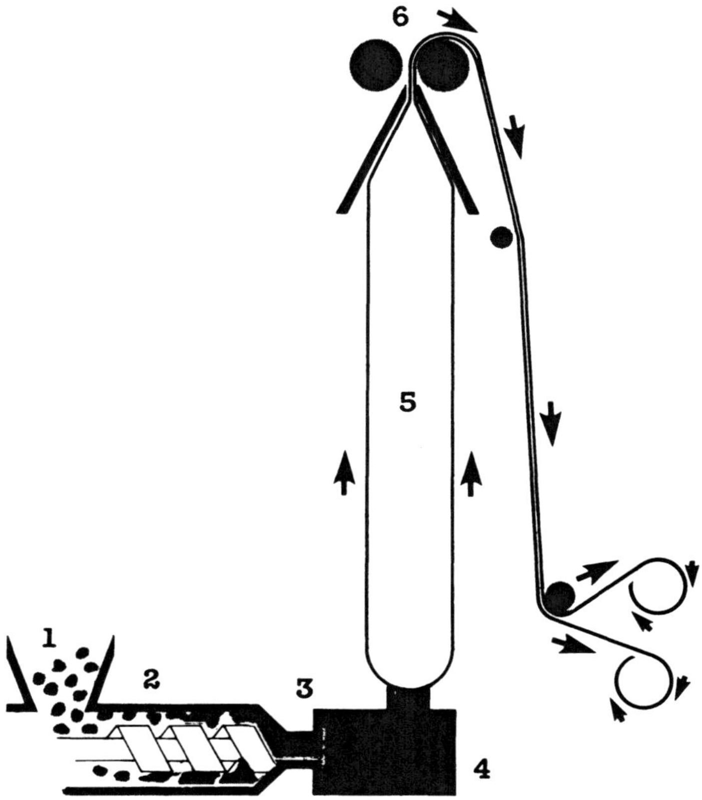

Figure 8.19 Method of manufacture of polythene. Blown film extrusion process: *1* Polythene film is manufactured from small granules of polythene. The granules are fed into a hopper which leads into a heated metal cylinder containing a large revolving screw; *2* As the granules pass along the screw the friction causes them to become a soft molten material; *3* At the end of the revolving screw there is a special section which makes sure that the melted plastic comes out at a steady continuous rate; *4* The molten polythene is pushed through a filter to remove any hard particles. It is then pushed through a hole called a 'die' which is in the shape of a ring; *5* Air is blown from the middle of the die to expand the polythene into a tube shaped bubble; *6* This is then pulled upwards through a pair of rollers and emerges as a flattened tube. The two edges of the flattened tube are cut off leaving two sheets of polythene which are separated and wound up individually; *7* If the film is to be printed it has to be treated so that the ink will stick. This is done with a high voltage electrical discharge, usually at the flattened tube stage; *8* Different coloured films can be made by adding coloured granules to the hopper at the start of the process. (courtesy CMB Packaging)

resistance at temperatures down to −50 °C, polystyrene is acceptable down to at least −40 °C and polypropylene copolymers down to about −20 °C.

Exposure to light may accelerate the destruction of chlorophyll in vegetables and the use of white-pigmented film or other means of creating opacity offers shelf-life advantages to such products.

The choice of materials/containers for packaging frozen foods therefore is

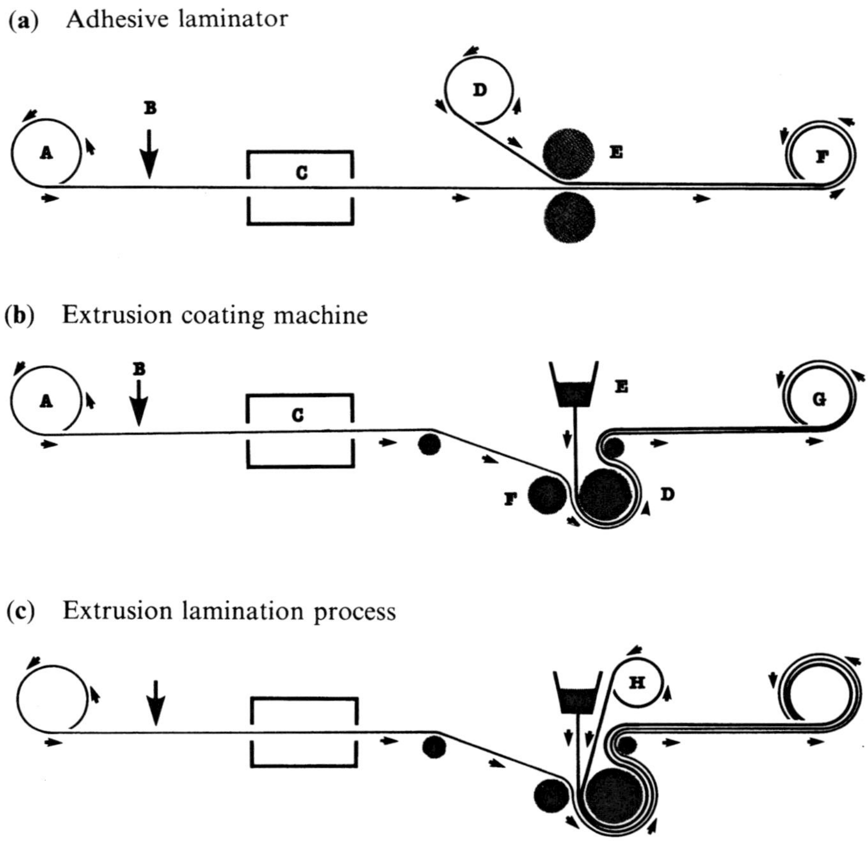

Figure 8.20 Methods of manufacture of laminates. (*a*) Adhesive laminator. The base film A (e.g. nylon) is coated with an adhesive solution B and passed through a drying oven C to remove all excess solvent. The base film is combined with the second film D (usually polythene), by passing both films through rollers E and the combined materials are rewound F. (*b*) Extrusion coating machine. The base film A is coated with a primer solution B to help the second plastic film stick. A drying oven C removes the solvent. The treated base film is pulled round a water cooled steel drum called the 'chill roll', D. Molten polythene E is extruded in the same way as in film production except the die is in the shape of a thin slot. The layer of molten polythene from the slot die E is extruded onto the primed base film, on the chill roll and the two layers are joined together by pressure from a rubber covered 'pressure' roller F. The laminate is then trimmed along the edges and wound up in reel form, G. (*c*) Extrusion lamination process. In this process, a third layer is added to the laminate by introducing a second film H, using extruded polythene as an adhesive. (courtesy CMB Packaging)

wide and is more dictated by product shape and economic/marketing factors than performance factors. The introduction of the plate freezer lent itself to the use of pre-filled cartons for freezing; cartons are still very widely used for frozen food packaging, the major advance in this field being the use of polythene-coated chipboard which can be filled and sealed at high speed, for example,

Diotite carton systems. Some carton-packed products, such as prepared vegetable dishes, require an inner pack such as a tray for support. With the increasing use of microwave ovens, the use of 'dual ovenable' (suitable for reheating in both conventional hot-air ovens and microwave ovens) or 'microwaveable' trays has increased enormously for frozen foods, these trays being of a variety of materials including polyester/board, coated foil or crystallised PET (dual ovenable) or almost any simple board or plastics material for microwave application. Some such trays are sleeve-wrapped rather than fully cartoned in order to reduce costs but maintain good presentation.

The individual quick freezing of particulate products, such as vegetables handled through a fluidised bed or immersion techniques, is now widely used and the products are packed into polythene bags bearing very high quality gravure printing. For those products intended to be cooked by the boil-in-the-bag technique, the most widely used packaging materials are nylon/medium density polythene, polyester/polythene and unsupported high-density polythene. Such packs are capable of being produced at high speeds on pouch form-fill-seal units.

8.5 Packaging of dehydrated vegetables

Dehydration is perhaps the most widely used method of food preservation; it is also one of the oldest. In general terms, dehydrated foods are more concentrated than other forms of foodstuff, they are less costly to produce, the minimum of labour is required, only a limited amount of equipment is used, storage requirements are at a minimum, distribution costs are low and packaging requirements are less critical than for some other processed foods. The chemical forces in dehydrated foods may be controlled by packaging and, in some instances, by certain chemical additives. Microbiological spoilage is basically controlled by reducing the free-water content below that necessary for growth. An additional pre-heat treatment (blanching) is considered necessary for vegetables prior to dehydration in order to inactivate enzymes, because dry heat has little effect on them. Enzyme activity is possible at moisture contents as low as 1%. It is difficult to define dehydrated foods in terms of moisture content, although many of the common dehydrated foods have moisture contents in the range 2–4%.

With dehydrated foods, packaging serves a multiprotection role. Depending on the type of product being packed, packaging has to provide protection against moisture pick-up, against loss of volatile components, against oxidation and physical protection against damage to the product itself. Some of the less sophisticated products simply require protection from moisture uptake to a greater or lesser extent and, for such products, a wide variety of containers and materials is available from polythene bags through composite and plastics containers to glass jars and metal containers with low-performance closures. Glass jars are, for example, very popular for spices and offer high presentational effect.

The trend towards higher quality dehydrated foods demands more protection than was necessary in the past. Many products, with moisture contents less than 2%, are highly hygroscopic and require total protection from the ingress of moisture. Such protection is given by hermetically sealed metal containers, hermetically sealed glass jars and heatsealed foil laminates in the form of sachets or pouches. Dried vegetable soups, for example, are frequently packed in paper/polythene/aluminium foil/polythene pouches on form-fill-seal machines, and such a material is also suitable for accelerated freeze-dried products, giving a shelf-life of several months.

Many dehydrated foods are also sensitive to oxygen uptake and loss of volatile components so that gas-impermeability is also required in the packaging material. A similar range of materials to the above will provide the necessary protection, perhaps with the addition of non-foil laminates incorporating PVDC. Simpler structures, used for example for dried onion rings and for some potato crisps, include PVDC-coated cellulose film (MXXT/A – where the cellulose film is coated with PVDC on both sides) or a cellulose film laminate such as MXDT/low-density polythene (where the cellulose film is coated with PVDC on only one side).

Apart from the necessity to protect the product from oxygen uptake through the package during storage, it may be necessary to remove as much oxygen as possible from within the pack at the time of filling. In-pack oxygen levels of less than 2% are required by a number of products to prevent fat rancidity. Sophisticated vacuum/gassing equipment, where the air in the pack is withdrawn and replaced by an inert gas such as nitrogen or carbon dioxide before final sealing has been developed to handle both rigid and flexible containers, and even higher speed units are planned; techniques for the vacuum/gassing of lined cartons (for example, Hermetet) and the vacuum packing of such containers are becoming well established or improved.

For those products that require physical protection from damage, a pouch-in-carton system, lined carton system (Hermetet), pillow pack or rigid metal/plastics container may be used.

8.6 Packaging of fresh/chilled vegetables

Fresh vegetables, by definition, continue to respire, namely they take in oxygen and give off carbon dioxide and water vapour. Some other gases, such as ethylene, are also produced. Fresh vegetables vary in their rate of respiration, with potatoes having a very low respiration rate and peas having the opposite. Their suitability for pre-packaging in film bags as fresh foods is indirectly related to their respiration rate.

Plastics materials for the packaging of fresh vegetables serve to reduce moisture loss and aid handling of the produce by the purchaser and, for most products, perforated low-density polythene is sufficient protection. The use of small perforations in the film for oxygen permeation has an insignificant effect on moisture loss – indeed, too good a moisture vapour barrier can lead to microbial spoilage. Other fresh produce may be supported in thermoformed PVC or expanded polystyrene punnets or trays, manually or mechanically overwrapped with thin low density polythene or ethylene-vinyl acetate cling films. Because it is possible to create a highly gas permeable version of it, plasticised PVC stretch films are also widely used.

When vegetables have been prepared in some way, for example, by peeling, slicing or shredding, they require to be handled as chilled foods, for which the recommended distribution temperature range is 0–8 °C. Although the deterioration of the product is retarded at such temperatures, their shelf-life may still be as short as 2–3 days unless changes are made to the composition of the gases surrounding the products (modified atmosphere packaging (MAP)). Once the modified atmosphere is created, the rate of change of that atmosphere will depend upon a number of factors, not least of which will be the ratio of the product to the internal volume of the package itself, the gas permeability of the packaging material used and the respiration rate of the vegetables.

It is necessary to control the carbon dioxide and oxygen and, with some systems, nitrogen levels in the pack. One approach is to use a gas-permeable film and modify the respiration rate of the product itself; an alternative is to use a relatively gas-impermeable film (e.g. PVDC/polythene) and incorporate carbon dioxide into the package before sealing.

Acknowledgements

The author wishes to thank colleagues at CMB Technology for their help and the company itself for permission to prepare this text.

Further reading

1. Hersom, A.C. and Hulland, E.D., *Canned Foods: Thermal Processing and Microbiology*, Churchill Livingstone, London, 1980.
2. Paine, F.A. and Paine, H.Y., *Handbook of Food Packaging*, Leonard Hill, London, 1983.
3. Ranken, M.D. (ed.), *Food Industries Manual* (22nd edn), Blackie, Glasgow, 1988.
4. Rees, J. and Bettison, J., (eds), *Processing and Packaging of Heat Preserved Foods*, Blackie, Glasgow, 1991.

9 Quality

DAVID LYON and HELEN CHURCHILL

9.1 Introduction

9.1.1 *Definition of quality*

The term 'quality' is one of the most defined terms in use in the food industry today. It is also perhaps one of the most confusing. It is obligatory, therefore, for authors on the subject to make clear to their readers which definition of quality they aim to take.

The British Standard (BS 4778: 1979) states that the word quality is used for several distinct purposes. Quality can be used in a 'comparative sense' whereby products are ranked on a relative basis or given a relative grade according to certain criteria. Alternatively, quality may be used in a 'quantitative sense', as is common in food manufacturing when discussing the acceptable quality level of a batch of product, or in other words the extent of departure from the ideal. The third, and perhaps most widely used concept of quality is the 'fitness for purpose sense' which relates to the ability of a product to satisfy a given need. This definition has been expanded by one author, who defines quality as the relative value of several characteristics of a product which determine the overall acceptability of the product to the buyer and ultimately the consumer (41).

This latter definition draws particular attention to the importance of the consumer in determining quality, and indeed, to the fact that overall quality cannot be determined without reference to the consumer and the use to which the consumer puts the product. A simpler definition, therefore, that quality is really about satisfying consumer needs, has its attractions and also perhaps some more thought provoking consequences for the processed vegetable manufacturer. Quality may therefore be defined as:

> '*The totality of features and characteristics of a product that bear on its ability to satisfy a given need*' (21).

This is the most widely used definition of quality and if considered closely can be seen to be made up of two parts. The first part, 'The totality of features and characteristics of a product...', concerns *objective* factors related to the product. The second part, '... to satisfy a given need', concerns *subjective* factors related to the user, or the consumer of the goods.

If we are to understand quality in processed vegetables, it follows that we must first understand the requirements of the consumer, and to consider what factors will influence consumers to purchase and eat vegetables. Clearly the intrinsic properties of the vegetables are unimportant if they fail to make the transfer from supermarket shelf to the dinner table. Having established the factors which influence consumer acceptance of their products, processors of vegetables must also be aware of how to react to these to ensure that they always produce foods that consumers want to buy.

9.1.2 *Lifestyle and trends*

In many western countries where food supply is plentiful and choice is almost unlimited, there continue to be many changes in the way people live. In the UK, the total population remains stable at around 56 million yet there have been considerable changes in the socio-economic structure of the country. These changes have had a significant influence on many aspects of everyday life, including daily eating patterns.

One major trend of great importance to the food manufacturer is the change in household structure. While the number of households continues to rise, this has been accompanied by a distinct drop in size with many more single-person households. There has also been a marked trend towards more married women taking full-time employment, giving them greater discretionary spending power and less time for food purchasing and preparation. Both these factors have contributed towards the greater use of convenience foods, to distinct changes in shopping patterns and to the trend for 'one stop' shopping.

The division of time for leisure and work, and the number of meals prepared and consumed in the home will also have inevitable consequences on the way people eat and what people eat. The tendency towards 'grazing' rather than established meal occasions and the phenomenal growth in snack-type foods which is predicted to continue well into the next decade (22), will greatly reduce the impact of conventional meals, and of the constituent foods within such meals.

There is also more awareness and greater interest amongst consumers in matters concerning 'natural, healthy' foods. Research has shown that the majority of consumers believe that vegetables are good for them and eating vegetables is more important for staying healthy than not smoking, taking regular exercise, being the right weight, getting enough sleep, and going easy on alcohol (23).

Manufacturers of processed vegetables can take some comfort, therefore, from the positive image that vegetables have in the eyes of the consumer. Indeed, it is forecast that the total market in the UK will increase from £3.2 billion to £3.8 billion within the next five years (25). A 14% increase in the sales of frozen vegetables and a rise in the sales of dried vegetables is forecast during the same period, as consumers seek convenience foods and as the proportion of individuals eating meat decreases.

Returning to the original definition that quality is about satisfying consumer needs, it is clear that conformance to lifestyle trends is a major quality factor for consumers. It is equally clear therefore, that manufacturers who recognise the need to develop products to fit such trends are most likely to benefit from the predicted increase in vegetable consumption.

9.1.3 *Consumers*

The needs of 'the consumer' have been discussed at length without defining what is meant by 'the consumer'. In fact, by referring to 'the consumer', it appears that there is a single body to be satisfied, ignoring the reality that the consuming population is made up of many individuals.

The fact that consumers act as individuals is not surprising when it is choice of fashion, the type of car driven, or the place lived in which is being considered. Equally, it is not surprising that consumers have different likes and dislikes in food. Pangborn (26) states that affective responses (degree of liking) are subject to many influencing factors, including previous exposure to the product, the method and context in which the food is eaten, and whether all the sensory cues are satisfied. It has also been argued that individuals determine their affective state by assessing cues which come from the environment (situational cues) and from the person's behaviour (self-produced cues). Furthermore, individuals differ in the relative extent to which they use one or the other of these two types of cues (27). Whether a consumer considers a vegetable to be of good or bad quality, therefore, depends as much on the environment under which the purchase and consumption decision is made, and the 'mood' of the individual, as it does the physical properties of the product. This highlights a difficult problem for the food manufacturer, as a product which is acceptable to an individual consumer on one occasion may be unacceptable to the same person on the next occasion, because it is out of context, or is required for a different use.

9.1.4 *Perceived quality*

While all the factors previously discussed are important in determining consumer perception of quality, there is little that a manufacturer can do other than respond to changes and trends. Manufacturers, however, can play a major part in ensuring that a customer continues to purchase their products, by controlling the characteristics which are considered important in good-quality products. This, in turn, requires an understanding of the factors which influence final quality and the methods used to measure and control it.

9.2 Factors influencing product quality

To produce high-quality products, the processor needs to be aware of the quality attributes which the consumer discerns as most important and which are most relevant in determining acceptability. Most consumers would initially judge the acceptability of products on their appearance, flavour, texture and perceived nutritional benefits. Each of these attributes is a function of the biochemical and physico-chemical composition of the vegetable, which is influenced by the quality and composition of the raw materials, the effects of processing and the effects of environmental factors, such as temperature, oxygen, light and moisture, encountered during storage and distribution, customer handling and use, and the barriers to these factors provided by the packaging.

9.2.1 *Selection of raw materials*

Generally, the characteristics of appearance, flavour and texture cannot be improved during subsequent processing, so the final quality of the product is heavily dependent on the choice of raw materials. Variations in raw material quality, particularly the critical quality parameters, may be crucial to the quality of the final product. It is important therefore to draw up and to adhere to detailed raw material specifications. These need to address the choice of raw materials and post-harvest handling and treatment, including storage and distribution procedures.

Vegetable processing has evolved from a way of preserving surplus produce into an industry geared to the efficient production of high-quality vegetables. The results of extensive work by plant breeders, genetic engineers, agronomists and technologists have all been used to grow raw materials most appropriate for the purpose of use. The importance of this work has been described in detail in an earlier chapter in this book.

From the point of view of final product quality, some factors are known to positively affect the sensory quality of vegetables after processing. Some pea cultivars, for example, which are used for freezing are considered unsuitable for canning as the skin tends to develop a bronze colour during processing and storage (12). The maturity most appropriate to processing must also be determined as this will have an effect on the texture of the final product. Other factors, important in raw material selection, may also indirectly influence final product quality. For example, by selecting bean cultivars with pods that are picked easily at harvest, leaves and strigs remain on the plant and result in reduced levels of extraneous vegetables matter in the final product. Equally, size and shape, which are important visual characteristics of products such as whole baby carrots and potatoes for canning, can, in some instances, be manipulated by agronomic practices in conjunction with cultivar selection.

9.2.2 *Post-harvest treatment and handling*

Vegetables continue to respire and carry on biological processes after harvesting. Bruising and internal damage which occurs after harvesting will clearly affect acceptability of vegetables, as well as the yield which remains suitable for processing.

Internal bruising may result from adverse conditions at the time of harvesting, handling or during post-harvest storage, even though there may be no visible signs of damage. Internal browning in Brussels sprouts and blackspot in potatoes (a blue-grey spherical zone in the vascular tissue which may take several hours or days to develop after minor damage) will affect perception of final product quality.

Disruption of internal structures of vegetables tissues, enables enzymic and chemical reactions to take place such that post-harvest delay in handling vegetables for processing allows deterioration of flavour, texture, colour and nutrients. In green beans, damage which occurs during mechanical harvesting may lead to the browning of the broken ends of the beans, and will present a severe problem if the raw material is held for extended periods prior to canning or freezing (18). Physical damage may also influence flavour. Peas have been found to develop marked off-flavours as a result of tissue damage, which stimulates the enzymic production of volatile off-flavours (13). Consequently, it is advisable to process as rapidly as possible after harvesting to minimise such changes. The alternative may necessitate storage under defined, controlled temperatures and humidity prior to processing in order to retain quality. For instance, peas held at higher temperatures have a firmer texture than those held at lower temperatures prior to processing (20).

9.2.3 *Processing and presentation*

Most vegetables are considered to be at their best quality when freshly picked and it is this quality that the vegetable processor is seeking to retain. After harvest, vegetables continue to respire and metabolise but in isolation from the rest of the plant. In time, the resources available for metabolism are depleted and the end products of metabolism accumulate. Many factors affect the rate of these reactions including intrinsic factors such as pH, water activity, the availability of enzymes and substrates and presence of inhibitors, and extrinsic factors such as temperature, light and the gaseous atmosphere around the vegetable. The rationale for most vegetable processing is to use one, or a combination of several of these factors to reduce or inhibit the rate of metabolic reactions and consequently retard ripening, senescence and the associated loss of quality and acceptability.

The preparation of retail packs of prepared vegetables for further processing by the consumer, for example, coleslaw and soup vegetables, has shown a marked increase in recent years (31). The vegetables in these packs are often chopped, diced, shredded or trimmed. Cut or shredded plant tissues deteriorate more rapidly and have a shorter shelf-life than produce marketed in a 'whole' condition. Small pieces of tissue are more susceptible to moisture loss and wilting, and cut surfaces dry out and frequently become discoloured. In addition, juices exuding from the damaged cells at the cut surfaces provide a growth medium for microorganisms. This can result in high numbers of microorganisms eventually leading to microbial spoilage and increasing the risk of growth of food-poisoning organisms (8).

All raw salad ingredients are inevitably contaminated with microorganisms, derived from either the growing environment or as contaminants during preparation. Although thorough washing can reduce numbers of organisms, the use of antimicrobial compounds such as chlorine will help to reduce the microbial load. Chemical antioxidants and free radical scavengers such as ethoxyquin and sodium benzoate, and chemicals such as sodium metabisulphite and ascorbic acid retard oxidation and discoloration reactions, as reported in work on potatoes (32) and mushrooms (34). A strict regime of chemical dips and treatments has also been shown to result in an acceptable shelf-life for carrots, cabbage, broccoli, cauliflower, celery, leeks, parsley and turnip (35). Furthermore, the use of sodium pyrophosphate/ascorbic acid treatment has been reported to eliminate the need for blanching of some vegetables prior to frozen storage (2).

Washing and peeling treatment prior to thermal processing or dipping may influence final quality. The quality of the water used throughout processing operations, including washing and recycled water, is important to guard against the introduction of taints which are carried through to the final product.

The method used for peeling can also influence quality. Steam peeling may induce some enzyme inactivation particularly in the outer layers of the vegetable. This may be responsible for the different observations reported by some workers on whether carrots require blanching prior to frozen storage. Schaller and Vamos-Vigyazo (16) reported no difference between blanched and unblanched carrot cubes after 18 months storage at −18 °C, whereas others reported off-flavours after 6 months at −20 °C (4).

Blanching is one of the most important operations prior to canning (see Chapter 4), dehydration, and freezing and confers many advantages to influence the final product quality and shelf-life. Texture, colour, flavour and nutritional quality during storage are stabilised to some extent by the destruction of enzymes at the time of blanching. Partial asepsis is conferred, and foreign materials such as pebbles, grit and berries which escaped during pre-cleaning are removed. Heat processes, and the leaching effect of the aqueous environment, are useful in reducing off-flavours such as pea-vining off-flavours and the levels of toxins and antidigestive factors such as haemagglutinins. The aqueous environment also enables the introduction of additives either via the blanching or cooling water. For example, it is reported that the colour of cauliflower is improved during subsequent storage if dipped in sulphite solution after blanching (14). However, legislation relating to residual sulphite may restrict this procedure. Addition of alkaline salts to neutralise acids formed during heat processing and storage is also a widely used method for preventing the formation of pheophytin, a breakdown product of chlorophyll which is responsible for the olive-green colour associated with heat-processed green vegetables. One of the main disadvantages of using alkaline salts is the possibility of inducing undesirable changes in texture. An alternative is to employ neutral salts, such as sodium chloride, which help to retain chlorophyll without causing changes in texture.

Canned vegetables present a particularly difficult problem with regard to pheophytin formation, and to prevent its development alkaline salts have been used at one or more stages of soaking, blanching and sterilisation. Blanching also removes air which is trapped around the fine hairs on the vegetable surface and expels air from between cells. This gives a brightening of the green colour of vegetables due to a change in the surface reflecting properties and prevents shrinkage and wilting of leafy vegetables. Without the latter, in the case of canning, the can would not be full after processing. Changes caused by the thermal process also confer consistency in processed products, for example, 'hot break' tomato juice, but the prolonged heating given to vegetables during conventional canning can also destroy plant pigments, resulting in colour changes of olive-green to brown-green, yellow-green and eventually yellow taking place. For dehydrated products changes induced in macromolecules at the time of heating ease the penetration of water at the rehydration stage (19). The quality of underblanched vegetables deteriorates rapidly during frozen storage and most vegetables require blanching to avoid quality changes. However, unblanched leek, onion and swede are reported not to change appreciably during 15 months storage at $-20\,^{\circ}C$ (4).

Other factors relating to blanching which influence the final quality of the product are the adequacy of the blanch and the method of blanching used. Factors which affect the adequacy of the blanch are related to the extent of enzyme inactivation, as both size and shape affect heat penetration into the vegetable or result in excessive heating in outer parts of the vegetable.

There are however some less desirable consequences associated with blanching, these include loss of flavour, colour, texture and nutrients. Marked differences in rate and extent of loss for each one of these attributes means that the processor has to make a compromise in selecting the operating conditions. Blanching renders vegetables more susceptible to infection by microorganisms and blanched vegetables should therefore be cooled rapidly to less than $10\,^{\circ}C$ to prevent rapid bacterial growth during the remainder of the preparation and packaging process. Blanching may also have an effect on pesticide residues, for example, ethylene bis dithiocarbamate fungicides (EBDC) are unstable in the presence of high moisture, temperature and oxygen. One of the breakdown products formed is ethylene thiourea (ETU), a compound believed to be teratogenic. Washing of vegetables prior to processing reduces further the levels of some pesticides. However, washing is not as effective for leaf-injured and wrinkled leaves that trap water and make residues more difficult to remove, for example, savoy and semi-savoy spinach (7).

Preservation of quality by freezing relies on the use of low temperatures (−12 to −18 °C) to prevent microbial spoilage and slow degenerative changes caused by chemical and biochemical reactions. The high water content of most vegetables (90–95%) gives rise to many of the problems associated with freezing. The rate of freezing influences the movement of water and the size of ice crystals that form. Fast freezing does not allow water to move before it freezes and many small ice crystals are formed, whereas slow freezing allows the movement of water resulting in the growth of fewer but larger ice crystals. In vegetables, ice crystal formation tends to begin outside of cells in intercellular spaces, leading to the movement of water out from the cell and concentration of remaining cell solutes. Disruption of cell structures and rupture of membranes by large ice crystals causes loss of turgor pressure which is important to the maintenance of a crisp and firm texture. Moreover, this disruption mixes enzymes and substrates that are usually separated by compartments within the cell, providing the potential for biochemical and chemical reactions leading to quality changes. For example, on freezing of raw green beans, bitter and rancid off-flavours develop during storage, although the volatile and non-volatile compounds which cause this have yet to be identified.

9.2.4 *Packaging*

Packaging used for commercially processed vegetables can contribute markedly to the retention of quality characteristics. The main function is to prevent contamination during handling and storage and, if contamination should occur, to prevent it from spreading to surrounding product.

The storage and distribution environment can be harsh, inflicting physical injury and dehydration, and initiating chemical and biochemical reactions that result in adverse changes in colour, texture, flavour and nutrition. Packaging technology can provide a barrier and some protection against such changes. Quick frozen vegetables are usually packaged in flexible, heat-sealed bags which are economical and easy to use. However, for brittle and fragile products such as cauliflower and broccoli, they provide little protection against physical damage. Aluminium foil laminates and cartons are alternatives which are being used to retain the quality of such products.

Loss of moisture from chilled vegetable products (as little as 5% by weight) results in shrivelling and wilting and increases the rate of senescence, for example in broccoli, yellowing and opening of florets is dependent on humidity. The use of non-perforated flexible film and flushing of the pack with modified atmospheres reduces the respiration rate and thereby slows deterioration. In frozen products moisture loss is seen as 'freezer burn', loss of colour and change in texture which are all usually associated with inadequate packaging. Moisture loss can be avoided by the use of moisture and vapour-proof packs.

Incident light can cause discoloration of product, initiate oxidation reactions and increase product temperature. Packaging may be used to protect against both specific wavelengths of light which may be important for initiating chemical and biochemical reactions, and the intensity of light which is responsible for the extent of radiant heat. Printing on the packet surface and the use of semi-reflective packs can assist in stopping the greenhouse effect, which raises product temperature within the pack.

The gas atmosphere contained within a pack affects the respiration rate and ripening and senescent processes in vegetables that have not received a thermal process. Harvested vegetables continue to respire and may exhibit a wide range of oxygen uptakes and carbon dioxide outputs. High respiration rates are usually associated with short storage life and preparation of the raw vegetable by slicing or shredding may cause an increase in respiration rate. Suitably permeable packaging films may permit the development of an equilibrium modified atmosphere after closing which results in slowing of respiration and retention of quality. In such packs, low levels of oxygen inhibit the growth of aerobic spoilage microorganisms and reduce the rate of oxidative changes but still allow respiration.

Oxidation in products, even when thermal processing has denatured enzymes, can be reduced by changes in the gas atmosphere. For products in which respiration is not a consideration, oxygen can be excluded by overwrapping with low oxygen permeable films or by evacuating non-permeable bags. Oxidative reactions that utilise oxygen from the headspace gases and oxygen held within the vegetable, can cause loss of colour and texture and lead to the formation of old and stale off-flavours during storage. These types of changes can occur in aseptically packed and HTST (high temperature–short time) vegetables in which enzymes have apparently been inactivated by the thermal process. During the storage of canned vegetables there is a small reduction in the volume of the headspace, indicative of this effect.

In some instances interactions of the product with packaging have been known to affect the colour and flavour of products. Blackening of canned green asparagus occurs shortly after opening the can. This is caused by the reaction of ferric iron from the can with the flavonol, rutin, which is present in high concentrations in green asparagus. It can be prevented if tin ions are present in excess. Under these conditions a soluble yellow complex is formed which does not discolour the product.

9.2.5 *Storage and distribution*

Storage conditions and the length of time in storage are often the most influential factors on the final quality. In chilled green beans, adverse storage temperatures affect the extent of changes in the nutritional value much more than cultivar and seasonal variations in cultivation conditions (29). Despite the importance of temperature, the processor may have little control over the conditions that a product will encounter once it has left his premises. However, it is necessary to have a knowledge of the storage temperatures and times likely to be spent at various stages of the distribution chain in order to prescribe a storage life, shelf-life or 'durability' for the product.

Temperature affects the chemical and biochemical reactions that are likely to occur, the microorganisms that are likely to grow, and the rate at which both of these will proceed. As a rough guide, the rate of chemical processes often double for every 10 °C rise in temperature. As time–temperature influences on quality losses are cumulative and irreversible over the storage life of a product, generally the higher the storage temperature, the shorter the storage life. For frozen vegetables this has been demonstrated in time–temperature tolerance (TTT) studies (36, 30). However, there are exceptions, some vegetables are 'chill sensitive', such that lowering the temperature to chill (1–8 °C) results in 'chill injury', seen as surface pitting, water rot, poor colour development and a general loss of structural integrity (11).

Temperature fluctuations and cycles are undesirable as the quality losses they cause are cumulative. The effect of temperature cycles is dependent on the sensitivity of quality attributes to changes in temperatures and how large and frequent the fluctuations in temperature are. The effect of temperature fluctuations on quality will also be dependent on the size of the pack and conductivity of the product. Temperature fluctuations for short periods may not have such a marked effect as slow fluctuations, as the product has insufficient time to undergo significant changes in temperature.

The influence that the gas atmosphere can have on quality and quality retention during storage has been illustrated by the use of vacuum and modified atmosphere packaging. When distributing or shipping mixed loads it is necessary to ensure that commodities are compatible with respect to temperature and humidity and the gases produced by all products, for example, lettuce is sensitive to low concentrations of ethylene such that mixing it with produce that emits substantial amounts of ethylene may result in russet spotting.

Incident light can affect product quality if it initiates chemical or biochemical reactions, for example the reactions involved in the synthesis of chlorophyll resulting in the greening of potatoes during storage, or oxidation reactions which accelerate the production of stale flavours and reduce the vitamin C content. Quality loss may also occur as a result of incident light of specific wavelengths, such as those which cause discoloration, or as a result of light intensity when radiant heat increases product temperature causing problems for temperature control in freezer and chill cabinets. Lighting cabinets with fibre optic probes rather than fluorescent tubes has been considered as a possibility for reducing radiant heat (5).

9.2.6 *Consumer handling and use*

Considerable effort is made by processors and retailers to ensure that foods reach consumers in the best possible condition. However, the way in which products are handled, stored and used in the home can affect quality, particularly for frozen and chilled foods which rely on temperature control for the maintenance of safety. In order to maintain quality up to the time of eating, it is in the interest of processors to educate and encourage consumers to improve and be aware of appropriate handling, storage and use of the product in the home. Fluctuations in temperature during 'carry home' time from retail outlet to home, followed by storage in domestic appliances designed for the maintenance of chill and frozen temperatures, rather than for chilling and freezing may result in adverse changes in quality. Thawing and refreezing is likely to result in changes in texture due to the formation of large ice crystals and dehydration of outer tissues of vegetables, and the formation of ice within the pack as a result of condensation formed during 'carry home'. If elevated temperatures are encountered, this may accelerate the subsequent rate of quality loss due to the growth of spoilage microorganisms and the action of enzymes.

Packs should be labelled clearly and concisely with instructions for storage and use. For example, the quality of vegetables remaining after the opening of a modified atmosphere package will not be retained for the duration of the shelf-life stated on the unopened pack. Cooking and other instructions such as defrosting are needed, as the extent of reheating or cooking required for processed vegetables prior to serving is as essential to quality as the original processing. Different methods of cooking such as boiling, steaming or by microwave will affect quality. Compounds which contribute to the flavour and aroma of cooked vegetables such as the sulphur volatiles in cabbage and the breakdown compounds of glucosinolates in brassica are influenced by the cooking time and method of cooking. Significantly higher levels of ascorbic acid are retained in vegetables on the plate after cooking in a microwave oven than in samples cooked by the most commonly used domestic method of boiling in water (9, 10).

It is not always easy, however, for the consumer to discern the extent of pre-cooking. The processor must, therefore, give clear and concise instructions for the consumer to follow. This is particularly important in those instances where microwaves are used for cooking or reheating. Instructions given by processors should have been validated in tests and should take account of the range of power ratings of microwaves found in customers' homes.

9.2.7 *Nutritional aspects – processed versus fresh*

It has been a widely held belief that processed vegetables have 'less nutrients' and are therefore inferior and not as 'good for you' as their fresh equivalents.

Much of the evidence for this, however, is based on research which has failed to consider the effects of typical post-harvest storage and domestic cooking on the composition of the fresh vegetable being used for the comparison. In addition, it has not always been recognised that vegetables for processing are usually harvested in prime condition and processed with minimum delay, while 'fresh' vegetables endure a variable distribution procedure until they are ultimately purchased by the consumer, and a subsequent variable storage period until they are actually consumed.

In a recent study, Hall and Edwards (24) have carried out a more realistic comparison of the compositional status of processed and fresh peas at the point, and in the state, at which they would be normally consumed. A single variety was grown on the same site, harvested at the appropriate maturity and separated into two samples to be treated as 'fresh' and 'processed' for the purposes of the study. The fresh samples were subsequently treated to typical conditions experienced in marketing, domestic storage and cooking while the processed samples were canned or frozen under simulated commercial conditions, stored and reheated as would be done in the home. The major effects of increasing maturity in peas were seen to be reduction in moisture content and vitamin C levels. Cooking in tap water resulted in the leaching of potassium and vitamin C and the uptake of calcium and sodium. The extent of change was dependent upon the amount of cooking liquor and the method of cooking, for example, conventional or microwave cooking. Reheating processed peas resulted in similar changes to those observed in cooking, but the compositional status generally remained stable over the storage period. The relative rates and extent of change in vitamin C content were such that the reheated processed peas were very similar to the fresh material. In fact, this research on peas together with similar studies on other vegetables showed that as consumed, processed vegetables are generally very similar in composition to fresh vegetables.

9.3 Measurement of product quality

One obvious way of measuring product quality is to monitor sales and customer complaints; the higher the sales and the fewer the complaints, the more likely one is to be satisfying consumer requirements. No responsible food manufacturer, however, would rely on this as their only method of quality control.

9.3.1 *Instrumental*

Quality assessments carried out using technical experts or trained sensory panels are disadvantaged by their requirement for manpower, and the perceived subjective nature of the results however carefully the conditions for carrying out the assessments are controlled. In addition, many vegetables are seasonal and results of quality assessments need to be reproducible so that comparisons from year to year can be made. Sensory panels cannot always provide precision from season to season, but instrumental methods of assessment can. The desirability of objective methods in developing standards for such products as peas, beans and Brussels sprouts has been recognised by the Codex Alimentarius Commission and the International Institute of Refrigeration.

The ideal instrumental method for measuring quality would be cheap, non-destructive to the sample, easy to use and not subject to variation or fatigue with a rapid response and wide application. All of these requirements are not satisfied by instruments currently available. In addition, to be successful, instrumental methods must be linked to assessments made by a trained sensory panel. Although there may be some cases where the sensory panel will be replaced by instruments, both will always be important as very often changes or differences in quality can be detected by taste, smell and visual examination before physical and chemical changes can be measured instrumentally.

9.3.1.1 *Colour*. Colour can be assessed more easily than taste, odour or texture. In the marketplace where the potential purchaser makes his selection he is seldom allowed to taste or eat food to test it; he can, however, look at it and judge by its appearance.

Foods appear coloured because the pigments they contain absorb light of some wavelengths and reflect or transmit light of others. Colour is affected not only by the concentration of pigments but also by the physical structure of the food and the way in which light is scattered from its surface. The definition of colour can be ambiguous. Scientists express colour as the abundance of neutral pigments for example chlorophyll, lycopene, carotenes and xanthopylls, whereas food technologists are interested in, for example, the degree of redness and yellowness shown by colour difference meters and that which the consumer can discern. Universally accepted general purpose methods for the objective specification of food appearance are not widely available. Total appearance analyses combining all colour and geometric attributes are too cumbersome for practical measurements or use. Instead, partial measurements are made of only those attributes which are important for each different product and problem.

Objective measures of colour attributes include spectrophotometric curves, CIE tristimulus measurements, colour difference measurements and measurements made on extracted colour pigments.

The spectrophotometric curve, determined using a reflectance spectrophotometer, is dependent on the form and light scattering properties of the sample. Assessment of colour is dependent on colour responses of the eye being combined with the measured reflectance curve. Although an accurate measure of colour can be achieved by this method, equipment is expensive and measurement slow. Simpler and more rapid methods that have been used extensively for quality control are CIE tristimulus measurements made using colour meters such as the Gardner or Hunterlab colour meters. This type of meter evaluates colour by means of three photocells which measure the hue (most dominantly reflected wavelength), saturation (intensity of colour) and luminosity (amount of light reflected from the surface) of a sample viewed in the reflected light from a standard source. The Gardner colour meter and Hunterlab colour meter both permit measurement on scales of whiteness to blackness, redness to greenness and yellowness to blueness. Combinations of these allow any colour to be measured. Alternatively, the same instruments can be used as colour difference meters. In this case the instruments are standardised against permanent colour standards (usually made in ceramic tiles) which, in colour characteristics, closely resemble the sample for examination. Instrumental and sensory assessments on peas, sliced green beans and Brussels sprouts have shown good correlations between sensory evaluations of 'green' and L- or b-values measured using the Gardner colour difference meter (1).

Other instruments for colour measurement are the Munsell discs and the Lovibond Tintometer, however these are not fully objective methods as they rely on the human eye for matching of sample colour to that produced by, respectively, combinations of rotating coloured discs or reference colours adjusted by an observer.

In the laboratory, the form and amount of pigment present in vegetable material and influencing the colour can be identified and measured (1). This requires the appropriate solvent mixture for extraction, separation using chromatographic techniques such as HPLC, and spectrophotometry to identify and quantify pigments by using their characteristic absorbance spectra. This type of fundamental approach might be used for trouble shooting or to investigate a specific problem rather than for routine quality control.

9.3.1.2 *Flavour*. The flavour of food is caused by chemical substances present in the food and consists of a combination of taste, odour and other sensations. Taste is caused by stimulation of taste buds on the tongue by soluble substances, nearly all of them non-volatile, released from the food into the saliva. Most of the contribution to flavour, however, is thought to be odour, due to volatile substances released from the food into the air in the mouth and carried into the nasal cavity. Although odour is usually described using simple terms such as 'green', 'mouldy', 'rotten', this belies the complexity of the chemistry involved. What is perceived is primarily dependent on the amounts of volatile substances that are released from the food and the way in which they interact to produce stimuli responsible for the perception of odour.

Vegetable flavour can be attributed to both volatile and non-volatile compounds. When eaten raw, vegetable flavour is markedly influenced by the volatile compounds that are generated and released as a result of enzyme activity during mastication, for example the characteristic flavour of raw cucumber is mainly due to C_9 unsaturated carbonyls. These are formed on chewing as the tissue disruption enables the concerted action of four enzymes to convert linolenic acid to unsaturated carbonyls (33). However, as a result of processing most enzyme activity is denatured and many volatile flavour compounds are lost through evaporation, so non-volatiles have a greater influence on flavour.

Measurement of flavours is either by chemical means or by the use of sensory panels. As instrumental methods are unable to take account of the interaction of the organism and stimuli to produce the perception of flavour, in most cases it is necessary to correlate measurements of particular components of a food with the intensity score or descriptors used by tasting panels for the same sample. The basic tastes, sweet, sour, salt and bitter can be measured simply when they are in pure form. For example, sugar solution can be measured by refractometers or in terms of degrees Brix, sourness can be measured using a pH meter and saltiness by the determination of chloride. The analysis of odour volatiles generally involves sampling of the gas from the headspace above the food, and in most instances a concentration of it by adsorption to a suitable support such as activated charcoal or an inert polymer, for example, Porapak Q. These extracted volatiles are then desorbed using either heat or an appropriate solvent prior to separation by chromatography, most commonly gas chromatography coupled to a mass spectrophotometer to identify the compounds present. However, it is a general problem in flavour chemistry that the progress in instrumental analysis has led to long lists of volatile compounds but not to the evaluation of their importance to flavour. Currently work is in progress towards the design and construction of an electronic device known as an 'electronic nose', which will be capable of mimicking the human olfactory system in discriminating with sensitivity between chemically complex odours (17). It is envisaged that this device will have applications for on-line discrimination of flavours.

9.3.1.3 *Texture.* The textural characteristics which consumers appear to value in vegetables are crispness and firmness. Texture is often the overriding quality factor in vegetables, such that, even if colour and flavour are acceptable, undesirable texture such as limpness, excessive hardness or softness, or mealiness causes the product to be rejected.

Factors that contribute to texture are the turgor pressure within cells which is important to crispness, and the strength of cell walls which, in combination with the adhesive properties that hold cells together provides rigidity, firmness and resistance to shear. During processing, damage to cell membranes by heat or ice crystal formation during freezing causes a loss of turgor pressure. Unless there is sufficient rigidity in the cell walls, this results in the collapse of tissues, for example if lettuce or tomatoes are frozen. In contrast vegetables with fibrous cell walls such as green beans are able to retain their form and are, therefore, more suited to processing.

Texture measurements are made to evaluate how foods will behave in response to handling, processing and storage, and to predict sensory properties which are related to texture and acceptance. Texture is an extremely complex attribute to measure and the approaches used require correlating measurements of texture to those described by sensory assessment. Light and electron microscopy have been used to observe changes in cell wall materials and adhesion between cells after different stages and forms of processing. Measurement of chemical composition such as alcohol insoluble solids is one of the most reliable methods for determining maturity of green peas although it does not relate accurately to tenderometer measurements, and specific gravity (starch or dry matter) of potatoes has been used as an indicator of mealiness and breakdown during cooking. However, these methods are time consuming. Different texture measuring devices are available which comprise penetrometers, compressimeters, shearing devices, cutting devices and masticometers (3). Some of these can be used for performing a number of different texture tests and provide flexibility and versatility, for example the Instron Universal Testing Machine. Within the processing industry simple and rapid measurements which indicate suitability for processing are required. The texture of peas and green beans are measured in both the USA and UK using either a tenderometer or a texture tester. The tenderometer measures shearing forces applied to a sample and its precision and accuracy have enabled maturity to be correlated with measurements of texture. For peas, measurements of maturity are used to indicate suitability for processing; tenderometer readings of 95 to 105 indicate maturity suitable for freezing and 115 to 120 are suitable for canning. The main disadvantage of the tenderometer is that it is not easily portable and cannot be standardised in the field. The texturemeter is less precise than the tenderometer but has the advantage of being a small instrument which can be transported relatively easily.

9.3.2 *Near infrared spectroscopy*

Near infrared spectroscopic (NIRS) analysis uses absorptions in the electromagnetic spectrum from 750 to 2500 nm which have previously been correlated to an analytical or quality factor of the food sample. Assessment of vegetable quality has included measurement of crop maturity of green beans, cooking time for processed peas (15), moisture, dry matter, colour, ripeness, sugars (6). The purpose of NIRS analysis is to obtain rapid and accurate analysis with a minimum of chemical expertise, often in conditions that would be unsuitable for chemical analysis. On-line methods using NIRS analysis are being developed and this is expected to be one of the most important applications for the future use of NIRS in the food industry.

9.3.3 *Immunoassay*

Enzyme-linked immunosorbent assays (ELISA) can be simple and inexpensive to perform requiring little expertise or equipment so they can be used away from the specialised laboratory, even in the field. These assays utilise enzyme labelled antibodies that are specific to the analyte in question. Detection is usually by the production of a coloured end product from a near colourless substrate. An increasing number of kits for performing analyses are becoming available. For example, antibody assays have been developed for vitamins (pantothenic acid, B_6 and B_{12}, and folate), contaminants (for example, mould in tomato paste), natural toxicants (including potato glycoalkaloids), groups of natural toxicants (including solanine and chaconine), glucosinolates (a group of compounds found in brassica which can form goitrogenic isothiocyanates and goitrin on hydrolysis), and more recently, for toxicants occurring as a result of human activity (such as herbicides and fungicides).

9.3.4 *Sensory evaluation*

Using instruments to predict sensory quality has long been the goal of many research workers, with the belief that such techniques will be without the human failings of subjectivity and fatigue. With modern food-processing factories working continuously, and with the high cost of labour, one can readily see the advantages in replacing as many human operators as possible with more predictable instruments. Equally, the enormous volume of business conducted with overseas suppliers, often without the advantage of a common language, would appear to be easier to control if requirements can be specified according to calibrated instruments.

The total sensory quality of a product, however, can only be predicted to a certain extent by instrumental, chemical and physical methods and the accuracy of these predictions can vary according to the product under test. As a consequence, Martens (28) has summarised the points in favour of using sensory analysis as the measure of product quality to be as follows:

(i) Sensory analysis cannot easily be replaced by other methods for quality evaluation.

(ii) Sensory analysis uses human senses as an instrument and can therefore measure many variables at one time.

(iii) Sensory analysis is more rapid and cheaper than many of the traditional chemical methods because, for example, colour, flavour and texture may be measured on the same sample.

(iv) Sensory analysis, used in the correct way, has no larger measurement noise than many of the traditional chemical analyses. In fact, since human beings are involved, there is an intelligent self-correction in a sensory panel that is often lacking in other instruments.

Clearly, the correct way to proceed is to take the advantages of sensory evaluation as a measurement technique and to integrate these with the advantages of appropriate instrumental techniques into a coordinated method for measuring and controlling quality.

One obstacle to the successful use of sensory evaluation by manufacturers has been apprehension with respect to setting up facilities, recruiting panels, selecting appropriate methods and analysing and interpreting data. Clearly, one must control the factors which would influence the measurement of quality attributes in vegetables, so that the results are both reliable and repeatable. This is as true for instruments as it is for sensory evaluation, and facilities providing the basic requirements are likely to be much easier to implement for sensory evaluation. All that is basically necessary, for instance, is a separate room with some control of the lighting and ventilation systems, and a simple booth construction to allow independent assessments to be made.

Selecting assessors for sensory evaluation must be carried out with as much care as is necessary in selecting any instrument used to measure food quality. Making the wrong decision is as important as purchasing the wrong equipment; it will simply be unreliable at best, or erroneous at worst. The selection and training of assessors is a wide topic to address in this chapter, as the type of selection procedure employed or the amount of training required, will depend to a large extent on the sensory task to be undertaken. For all types of tests, however, it is important that the assessor is well motivated and has the correct attitude towards the test. Poorly motivated assessors quickly lose interest in their work, become hasty, careless and apparently poor at discrimination. They can also start to influence others leading to a general despondency in the panel. It is important, therefore, to ensure that assessors are aware of their contribution to the success of maintaining and improving quality of products, and to be aware that their employers have faith to act on their recommendations and observations.

The approach used by assessors is of paramount importance. All observations should be independent, objective and be made by individuals who are confident in their own judgement. Domineering assessors who may seek to impose their views on others should not be selected. For the same reasons, it is always wise to avoid a mix of senior and junior staff, and to exclude assessors with any particular bias. It can be argued that in certain situations, for instance, production managers may be more influenced on questions of volume than questions of quality.

There are several sensory methods which can be used to measure the quality of vegetables during production and supply. The use of product specifications in the manufacture, purchase and supply of food items is perhaps the most widespread method used in normal commercial practice. Many styles of specification exist but the principal factor which distinguishes a well-prepared specification from a poor one, is whether the method effectively combines the objective factors related to the product with the subjective factors related to the user.

This relationship between objectivity and subjectivity has to be clearly understood if specifications are to be used reliably to measure product quality. Examples of objective factors related to the product are presence of defects, the colour, flavour, texture of the product and the size of individual units. Subjective factors related to the user will include whether a defect is considered objectionable, whether the flavour is acceptable or whether the user prefers regular or irregular sized units.

9.4 Product specifications

9.4.1 *Deriving specifications*

There are several important considerations to be made in putting together a quality specification and these can be listed as follows;

1. *The requirements must be technically and economically feasible.*

A quality specification should be an effective tool for measuring product quality and for communicating product requirements from the purchaser to the supplier. It is vital, therefore, that the requirements laid down by the purchaser are practical, and achievable at the price that the purchaser is prepared to pay. This may inevitably lead to compromise on the part of the user.

A typical example to illustrate this point is size grading of whole carrots for freezing. Carrots are normally graded by crown diameter by passing them over diverging bars, but this method of grading takes no account of carrot length. If it is important to specify carrot length, the user should be aware that although technically feasible, this would represent a diversion from normal practice and would require an additional handling procedure. This may also involve the purchase of additional grading equipment and almost certainly would result in higher prices for the end product.

2. *All statements must be unambiguous.*

If a specification is to work, every statement must be clear, concise and entirely unambiguous. Ambiguity causes confusion and inevitably leads to areas of dispute between parties when products fail to achieve the required standard. An example of an ambiguous statement can be found in the Codex Alimentarius Commission Recommended International Standard For Canned Tomatoes (38) which reads as follows:

> 2.3.5 *Defects*
> The finished product shall be prepared from such materials and under such practices that it shall be substantially free from objectionable core material

and extraneous plant material and *shall not contain excessive defects whether specifically mentioned in this standard or not.*

Clearly the purchaser would feel it within his right to reject a product containing anything he did not like or personally found 'excessive', irrespective of the nature or the level of the defect or the opinion and the views of the supplier. Not only is this statement ambiguous, it also fails to differentiate clearly between product factors and user requirements.

3. *The specification must clearly define the characteristics to be measured.*

While a statement may not be ambiguous, it may be phrased in such a way that misinterpretation occurs more often than it ought to. The comparison of the two statements defining dice size in frozen or canned carrots listed below clearly illustrates this point.

(a) Campden Specification for Quick Frozen Diced Root Vegetables (39). Oversize and undersize dice – nomial 10 × 10 × 10 mm dice:

> Oversize dice are larger than 15 mm in any one or more dimensions. Undersized dice are any dice less than one third of the nominal size in one or more dimensions. Undersize pieces are those which fall through a sieve having square holes 8 × 8 mm.

(b) USDA Specification for Canned Carrots (40). Uniformity of size and shape – diced carrots:

> The units are practically uniform in size and shape with edges, other than the rounded outer edges, measuring approximately $\frac{1}{2}$ inch or less; and the aggregate weight of all units of irregular shape which are noticeably smaller than one half the volume of an average size cube and of all noticeably large and large irregular shaped units does not exceed 12% of the weight of all the units.

In (b) the definition is rather cumbersome and difficult to interpret. This introduces the possibility of error and increases the time required, perhaps disproportionately, to evaluate this characteristic.

4. *A specification must clearly state the allowable tolerances on product characteristics and the degree of flexibility permitted.*

If the quality specification is to be used as a tool for quality control during production, it is important that the operator can clearly see when and why products are going out of specification. This will allow prompt action and feedback to try to rectify the situation. Tabulated tolerance for defects allows quick reference of this kind, but of course this must also be accompanied with clear written definitions of the defects in question.

Sampling rates are also important. It is clearly impractical to carry out 100% sampling of food items, whether during production or during quality checks by the purchaser. It is equally impractical to expect 100% of the samples checked to be within the specified limits, particularly when dealing with biologically variable material. A good specification would therefore indicate the sampling regimes to be used, usually by reference to accepted sampling procedures, and the AQL (acceptable quality level) to which these sampling procedures should be applied.

5. *The specification must properly address the sensory aspects of the product.*

This part of the specification often receives less attention than it deserves, and is sometimes thought of as the least important part. It is also the easiest aspect to criticise as being too subjective and non-reproducible.

The importance of satisfying the needs of the user has already been covered and discussed as an integral part of the definition of a quality specification. If food looks good and has a good flavour and texture, not only would it be more acceptable to the consumer, but would add to the enjoyment of the whole meal. Measuring and controlling the sensory aspects within a specification must therefore be of great importance.

Equally, there are factors which cannot be determined effectively without the use of sensory evaluation. A taint, for instance, can only be detected by people. Instruments merely measure the level and nature of the chemical responsible for causing a foreign flavour or odour in the product.

On the question of subjectivity, it has been proved on many occasions that people can become objective and descriptive in their evaluation of food. This requires experience and training, and such a panel or group of individuals can become extremely sensitive and their results consistently reproducible.

6. *The specification must be agreed between the supplier and the user.*

Although this appears to be an obvious necessity, it is not unusual for products to be bought without reference to a product specification. In these particular cases, it clearly cannot be subsequently established whether a product satisfies the agreement made between the parties involved. In cases of dispute, this of course makes the settlement of the dispute even more difficult. Where product specifications are used, often they are not always fully understood by both parties. A purchaser must make sure that he is fully aware of what he is asking the manufacturer to supply, and the manufacturer fully aware of what he is expected to deliver.

Quality specifications are best derived and agreed by the people who wish to use them. They should be clear, concise and, above all, they should be unambiguous. They should be interpreted by trained personnel in an objective manner, and they should clearly combine the factors related to the product with the requirements of the user. If they achieve this, they will form a very effective tool in specifying and monitoring product quality.

9.4.2 *Types of specification*

9.4.2.1 *Voluntary specifications.* As previously discussed, a specification is a method which quantifies product quality by combining the objective factors related to the product with the subjective factors related to the user. It is clear that someone has to make a decision as to what is acceptable and what is not acceptable. In the absence of data directly generated from consumers, one must rely largely on the experience of product specialists to convey what they believe consumers want. This must be listed clearly and precisely, and subsequent interpretations carried out objectively by trained personnel, to ensure that the element of subjectivity is removed.

Campden Food and Drink Research Association (UK) has written specifications for a wide range of canned and frozen vegetable products (for example, quick frozen diced root vegetables (39)). These are derived through industrial committees with representation from growers, manufacturers, wholesalers, importers, distributors, retailers, caterers and staff at the Research Association. A discussion document or first draft is produced by Campden in the style and format agreed for previous documents, by reference to any documents or standards supplied by members of the committee. This first draft document will then enter a series of draft stages at meetings of the committee, until an interim specification is produced. Most of the major alterations to the specifications are carried out during the drafting, so the interim document is intended to be very close to the final specification. Its purpose however, is to ensure that the standard is fully applicable in commerce before it is issued as a final specification.

Campden specifications are used by the food industry as trading documents on which contracts are based. This is the primary use for any product specification, but as the Campden specification is a voluntarily adopted industry standard, it has a particular advantage for the supplier. A manufacturer, for instance, is likely to be in a better position to satisfy his many customers with one common standard, than he is with a different requirement from each of them.

If a specification is used as a trading document, it follows that it is also used in quality control by the manufacturer and in quality assurance by the purchaser.

9.4.1.2 *Government-backed specifications.* In the UK, there are no statutory quality grading systems for processed vegetables. All quality specifications are agreed independently between purchaser and supplier, perhaps using Campden specifications as a basis for this agreement. In other countries, however, standards are drawn up by government departments and sampling to check conformance to specification carried out by government inspectors. Probably the most important standards of this type are those developed by the Federal Agency and the United States Department of Agriculture (USDA) (40).

Processed vegetable standards are derived from the basic requirements laid down in the Federal Food, Drug and Cosmetic Act 1938. Federal Standards are published in the Federal Register and these standards are mandatory. Product which conforms to the Federal Standards can also be graded according to the USDA specifications. These specifications normally classify product into four grades A to D, often referred to as US Fancy, US Choice, US Standard and US Substandard. Accompanying these specifications are detailed sampling procedures and regulations for inspection by USDA appointed inspectors. Product thus inspected is allowed to be marked with an official mark stating the grade and the fact that the product was sampled or packed under the continuous supervision of the USDA.

Similar government-backed standards for monitoring product quality exist in other countries, and in the early 1960s, the Codex Alimentarius Commission was set up to coordinate the work of several organisations involved in setting standards, to provide an effective means by which national governments could accept these standards. The intention behind the Codex initiative was to harmonise the requirement for foods, as much as was possible, and in so doing to make international trade easier while maintaining the interests of the consumer. Within the European Community, steps have been taken to introduce common and agreed specifications for processed vegetables but these were not pursued, and there are currently no plans to do so.

References

1. Adams, J.B., Robertson, A. and Springett, M.B., Campden Food Preservation Research Association Technical Memorandum 307, 1982.
2. Anon, *Food Engng* (September 1988), 92.
3. Armstrong, K.W., Springett, M.B. and Adams, J.B., Campden Food Preservation Research Association, Technical Memorandum 228, 1979.
4. Baardseth, P., *Food Chem.* **3** (1978), 271.
5. Birnie, W., *Frozen and Chilled Foods* (October 1989), 25.
6. Davies, A.M.C. and Grant, A., *Int. J. Food Sci. Technol.* **22** (1987), 191.
7. Gonzalez, A.R., Davies, D.R., Katten A.A., Elkins, E.R. and Kim, E.S., *J. Food Safety* **9** (3) (1989), 155.
8. Geeson, J. and Brocklehurst, T., *Food Process.* **58** (4) (1989), 41.
9. Gould, M.F. and Golledge, D., *Food Sci. Nutrit.* **42F** (1989), 145.
10. Hall, M.N., Edwards, M.C., Murphy, M.C. and Pither, R., Campden Food and Drink Research Association Technical Memorandum 553 1989.
11. Jackman, R.L., Yada, R.Y., Marangoni, A., Parkin, K.L. and Stanley, D.W., *J. Food Qual.* **11** (1988), 253.
12. Lee, C.Y., in *Quality and Preservation of Vegetables* (ed. N.A.M. Eskin), CRC Press Inc., Boca Raton, Florida, 1989, 159.
13. Murray, K.E., Shipton, J., Whitefield, F.B. and Last, J.H., *Sci. Food Agric.* **27** (1976), 1093.
14. Ramaswamy, H.S. and Ranganna, S., *J. Food Qual.* **11** (6) (1989), 443.
15. Scotter, C.N.G. and Hall, M.N., Campden Food and Drink Research Association Technical Memorandum 511, 1988.
16. Schaller, A. and Vamos-Vigyazo, L., *Lebensmittel- u. Biotechnologie* **3** (1986), 143.
17. Shurmer, H., *Phys. Technol.* **18** (1987), 170.
18. Sistrunk, W.A., Gonzalez, A.R. and Moore, K.J., in *Qaulity and Preservation of Vegetables* (ed. N.A.M. Eskin), CRC Press Inc., Boca Raton, Florida, 1989, p. 185.

19. von Loesecke, H.W., in *Drying and Dehydrated Foods*, Reinhold Publishing Co., New York, 1965.
20. Weckel, K.G., Seeman, B. and Ng, W.C., *Food Technol.* **18** (8) (1964), 97.
21. European Organisation for Quality Control, *Glossary of Terms Used in Quality Control* (4th edn), 1976.
22. Sills-Levy, E., *Food Technol.* **43** (1989), 128–132.
23. Salmon, J., *Acta Horticulturae* **244** (1989), 35–43.
24. Hall, M.N. and Edwards, M.C., *Acta Horticulturae* **244** (1989), 217–226.
25. Mintel Market Intelligence Report, *Canned Fruits and Vegetables*, 1988.
26. Pangborn, R.M., in *Criteria of Food Acceptance*, Forster Publishing Ltd, Zurich, 1981.
27. Stevens, D.A., Dooley, D.A. and Laird, J.D., in *Food Acceptability*, Elsevier Science Publishers Ltd, England, 1988.
28. Martens, M., Dr.Agric. Thesis, Agricultural Univ. of Norway, 1986.
29. Bottcher, H., *Die Nahrung* **30** (9) (1986), 921.
30. Lyon, D.H., McEwan, J.A., Taylor, J.M. and Reynolds, M.A., *Food Qual. Pref.* **1** (1) (1988), 37.
31. Mintel (UK) Special Report, 1985.
32. O'Beirne, D. and Ballantyne, A., *Int. J. Food Sci. Technol.* **22** (1987), 515.
33. Philips, D.R., Matthew, J.A., Reynolds, J. and Fenwick, G.R., *Phytochem.* **18** (1979), 401.
34. Saxena, S. and Rai, R.D., *Mushroom Journal for the Tropics* **8** (1) (1988), 15.
35. Strugnell, C., *Irish J. Food Sci. Technol.* **12** (1 & 2) (1988), 81.
36. WRRL. Cited by Talburt, W.F. and Smith, O., *Potato Processing*, AVI Publishing Co., Connecticut, 1975.
37. British Standard No. 4778, 1979.
38. Codex Alimentarius Commission Standard for Canned Tomatoes. Joint FAO/WHO Food Standards Programme, Rome, 1969.
39. Campden Food Specification Quick Frozen Diced Root Vegetables, Campden Food and Drink Research Association, 1988.
40. *Code of Federal Regulations*, US Government Printing Office, Washington, 1988.
41. Arthey, V.D., *Quality of Horticultural Products*, Butterworths, 1975.

10 Plant, equipment and cleaning

DAVID TIMPERLEY

10.1 Introduction

The location, the site, the design and construction of the building, its layout, the equipment, cleaning system, water supply and effluent disposal are among the many factors which have to be considered before establishing a factory.

10.2 The location

A major requirement for the processing of many vegetables is that the factory must be in the proximity of the growing area, not only to minimise transport costs but because some harvested crops such as vined peas, have to be processed within a short period. It is important that an area is chosen in which crops can be grown all the year round to ensure that full utilisation of the plant can be made for optimum profitability. Some crops, such as peas, have a season as short as six weeks, necessitating a twenty-four hour day processing operation, whereas the season for processing other crops, such as Brussels sprouts, may be as long as three months. Other factors to be considered must be the proximity to good transport systems, power supplies, the availability of labour and probably, most important of all to a vegetable-processing factory, a good supply of potable water, as well as adequate effluent disposal.

10.3 The site

Careful consideration must be given to the selection of the site to avoid possible problems. The site itself must be large enough, not only for present requirements but for possible future expansion. There needs to be sufficient space for fairly large areas of hard standings for vehicles delivering raw materials. In addition, due to the possible high levels of organic material in the waste water which results from most vegetable processing operations, it may be necessary for the factory to have its own treatment plant for which space would be required. Detailed inspection should be made of any surrounding industries or activities which may emit smoke, dust, odours or other contamination. Reclaimed land, especially that which has been in-filled with refuse, should be avoided, as should sites from which pests and insects cannot be excluded and areas which are not free from flooding.

10.4 The building

Existing buildings which have previously been used for other purposes will have to be considered most carefully for their suitability. It may not, for example, be possible to decontaminate them or the cost of refurbishment may be prohibitively high; there will, of course, be no such problems with a 'green field' site. In general, single-storey buildings are favoured because it is easier to ensure straight-line processing as well as ease of movement of personnel and products, maintenance and sanitation. It is, additionally, easier to rearrange equipment to suit the processing of different crops throughout the year. Initial processing operations such as the removal of extraneous vegetable matter (EVM) by penumatic separation or the initial washing of root crops may be carried out in the open or in open-sided buildings. As the product flow proceeds, then increasingly higher standards for the building are required.

There is little guidance from any present legislation relating to internal food-processing factory finishes. In the United Kingdom the Food Hygiene (General) Regulations 1970 are the principle regulations. These state that the walls, floors, doors, windows, ceiling, woodwork and all other parts of the structure of every food room shall be kept clean and shall be kept in such good order, repair and condition as to enable them to be effectively cleaned and prevent, so far as is reasonably practical, any risk of infestation by rats, mice or insects.

The floor is considered to be one of the most important parts of the building and is worthy of high initial capital investment because disruption of production caused by the repair of defects can be very costly. Whilst the selection of the type of flooring will depend upon the process being carried out the floor should be impermeable and easily cleaned. Impermeability and slip resistance are properties which are obviously very important in areas where there are large quantities of water on the floor. Concrete is not considered to be suitable as the final finish because it is permeable and is soon eroded in areas exposed to large quantities of water. This exposes the aggregate and makes cleaning difficult. In dry areas where, for example, packaging operations take place, the problem with concrete is that of dust due to its porosity.

Fully vitrified ceramic tiles having an antislip top surface will, if properly laid on a sound base and suitably grouted, be satisfactory for many years. A large range of finishes is available such as seamless resin screeds which are applied as a topping to concrete. Floors should have a fall to drain of between 1 in 50 and 1 in 100 depending upon the process operation and the surface texture and laid such that 'puddling' is avoided. A compromise for general purposes and safety is 1 in 80 as proposed by Cattell (1). The distance between drains should be no more than about 5 m according to Hayes (2). The type of drain will depend largely upon the process operation; in very wet areas open gutter drains or channels are most suitable, especially if vegetable matter is present. Katsuyama and Strachan (3) and Imholte (4) suggest that drainage channels should be about 150 mm deep (if they are deeper then cleaning is difficult), have smooth vertical sides and rounded bottoms with a fall of at least 1 in 100. Should gratings have to be fitted for safety reasons or to allow the passage of wheeled traffic, then the top edges of the channel should be so designed and constructed to avoid damage. The size of the openings of the slots or holes in the gratings should be such that they do not become clogged by solids. Discharges from equipment should, where possible, be fed directly into drains to avoid situations in which the discharge from a blancher has to flow across the floor to the drainage channel. A solution to this problem is to build a low wall around the equipment from which water and solids may be drained, as shown in Figure 10.1. Where the channels are close to a wall they should not be directly against it to avoid flooding of the wall to floor junction (5). An indirect advantage of channels near a wall is that the siting of equipment hard up to the wall is prevented, thus providing access for cleaning. The drainage system should flow in the reverse direction of production and solids must be separated from liquids as soon as possible, by screening, to avoid leaching and subsequent high effluent concentrations (6). Traps should be easily accessible, frequently emptied and outside the processing area. The floor should be coved where it meets walls or other vertical surfaces such as plinths or columns as this facilitates cleaning.

Walls should be smooth, durable, impermeable and easily cleaned; a wide range of finishes is available to meet such requirements. Solid walls may be

Figure 10.1 Containment of discharges of water and solids by a low wall.

tiled, coated or clad. Fully bedded, properly grouted, ceramic coatings may be applied to fair-faced concrete, rendered surfaces, blockwork, and brickwork, and can provide an impermeable finish. Cleanability is, of course, affected by the surface roughness, for example, coated blockwork will not be as easily cleaned as coated render. In very wet or humid areas, where there is a strong possibility of mould growth, the application of a fungicidal coating may be considered; there is evidence that such coatings remain effective for many years. Cladding of walls with thermoplastic sheet is another finish which is widely used in refurbishment. As with any system involving joints special attention must not only be paid to that between panels but also to the wall/floor and wall/ceiling juncture. In some areas there is a need for insulated panels to be used. Insulation which may be up to 200 mm thick is sandwiched between and bonded to coated sheet steel. Again, careful attention must be given to the jointing system; if panels are used as partitions and not as cladding, then the floor joint can be a problem. One solution is to fix and seal the panels to a low wall coved to the floor. All walls should be protected from damage where it is likely to occur, for example, trucking routes. Kerbs or crash barriers, preferably supported from the floor, should be installed. Careful consideration must be given, not only to the fire retardation of the wall coating material, but also to the toxicity of the fumes emitted in the event of a fire as these could hamper a fire-fighting operation.

Windows should be glazed with either polycarbonate or laminated glass and, if opening, protected by insect screens. Internal sills should be sloped, at an angle of at least 45°, to facilitate cleaning and, when at a low level, to prevent them from being used as shelves. Most modern factories are windowless, hence avoiding any problems associated with glass, ingress of insects or fumes, and ensuring even, shadow-free illumination from artificial lighting.

Doors should be kept to a minimum and those for the access of personnel should be self-closing; emergency exit doors must be fitted with 'out only' operating bars. Larger doors which have to be open for longer periods, if used for trucking for example, should be fitted with properly designed air curtains to prevent the entry of vermin. Metal or glass-reinforced plastic (GRP) should be used as the material of construction as wood is more easily damaged and requires more maintenance; door frames should, preferably, be of metal.

The ceiling should be smooth, uncluttered and easily cleaned. Figure 10.2 illustrates the open type structure which is typical of many installations. In areas prone to high humidity, materials and finish must be selected to minimise flaking and mould growth. The combination of both natural and artificial lighting is shown in Figure 10.2. Artificial lighting has many advantages in that, if properly arranged, it provides even illumination over inspection belts. Fluorescent tubes and lamps must be protected by shields, usually of polycarbonate, to protect the glass and contain it in the event of breakage. Suspended units should be smooth, easily cleanable and designed to

Figure 10.2 Typical processing area.

the appropriate standards to prevent the ingress of water. It is suggested that lighting units are plugged in so that in the event of a failure the entire unit can be replaced and the faulty one removed from the processing area to a designated workshop for maintenance. Ideally, recessed lighting flush with the ceiling is recommended from the hygienic aspect but this is not always possible and maintenance may be difficult.

Adequate ventilation is particularly important for most vegetable-processing operations because of the large quantities of steam and moist air involved. It is necessary, not only for the comfort of personnel but also to reduce the possibility of mould growth, as well as to avoid corrosion and attack on internal finishes. Canopies and extraction systems should be provided above certain equipment to ensure that problems are dealt with at source. Air inlets to processing areas should be suitably screened to prevent the ingress of insects but must be accessible for regular cleaning. It may be necessary in chilled vegetable production areas to install microbiological air inlet filters and to ensure that there is a slight overpressure in the area. In general, the flow of air should be in the opposite direction to the production flow. If frying operations take place in, for example, the case of potato products, then removal of the odours may be necessary by means of an activated carbon absorption system or by combustion before discharge into the atmosphere.

Services within processing areas, such as pipework, should be installed in such a way that they do not constitute a hygiene hazard. No service pipes should pass over product lines so as to avoid the possibility of contamination by leakage or condensation. Pipework is usually supported by wall-mounted brackets or suspended from ceilings; it must be arranged so that it can be cleaned by having adequate clearance with walls or ceilings. Pipe supports and hangers should be of simple construction and easy to clean. Jacketing of insulated pipes by cloth or canvas is unacceptable as this can harbour insects and allow mould growth. Insulated pipework within processing areas must, therefore, be clad in metal or rigid plastic having sealed, welded or bonded joints. It must also be realised that ladders will probably be placed against pipework on occasions and the cladding must be able to withstand possible mechanical damage.

The support of electrical supply cables to processing equipment is one of the greatest hygiene problems because so little attention is paid to it during the design and installation stages. Perforated cable trays are so often to be seen within processing areas. Cables may be run in trunking or conduit but this must be effectively sealed against the ingress of vermin and water. It must be supported such that there is adequate clearance for cleaning. It is sometimes possible to carry electrical wiring through the tubular sections of the framework of the equipment but such systems have to be given careful consideration at the design stage and require very close liaison between the mechanical and electrical designers (7). All switchgear and controls, other than emergency stop buttons, should, whenever possible, be sited in separate rooms away from processing areas, particularly if wet operations are taking place.

The water supply is of particular importance in most vegetable processing operations because of the very large quantities required for both washing the products and for sanitation. In both cases the water must be potable and tested periodically to ensure that it is safe and unpolluted. As most of the water-transmitted diseases are caused by bacteria from the human intestine a test for coliform bacteria should always be made. In addition, water should be checked for excessively high quantities of calcium ions which will harden vegetables, or nitrate ions which cause de-tinning and corrosion of containers which are tinplated. In order to economise on water usage water is often recirculated or re-used in some stages of processing, such as cooling or washing, where the contraflow principle is used, that is, incoming dirty product is washed by used water which is then dumped. In such cases it is necessary to rechlorinate the water to ensure that it is microbiologically safe. Soil is a large reservoir of bacterial spores and root crops and dwarf-growing plants usually carry large numbers of microorganisms after harvesting, particularly in wet-growing and harvesting conditions. Peas harvested by mobile viners are particularly prone to heavy bacterial contamination and tomatoes are another crop which also can suffer from heavy bacterial contamination as the result of soil being splashed over them by summer showers (8). Careful attention must, therefore, be paid to the cleaning and washing operations if unacceptable levels of contamination are to be avoided. Hulland (8) does not recommend the use of recirculated water for the washing of vegetables because of the danger of rapid build-up of microorganisms in the system. Chlorination of recirculated water would be largely ineffective because the free chlorine would be absorbed rapidly and neutralised by the organic content of the water. Should it be necessary to store water, then care should be taken to ensure that it is protected from contamination by holding it in closed, non-corrosive vessels and is chlorinated before use. Water supply pipelines should be free from 'dead legs' which could allow stagnation and an increase in the number of microorganisms. Non-potable water is used for some purposes such as to supply fire hoses or cooling towers but as it is a potential hazard to the product it must not be used for any other purpose, such as sanitation and must be identified and not connected in any way to the potable water system.

Considerable quantities of steam are used in many vegetable processing operations such as peeling, blanching and cooking. Any steam coming into direct contact with the product must not contain any additives which may result in contamination.

Effluent from vegetable-processing operations tends to be very high in biological oxygen demand (BOD); hence, it is important to have a satisfactory effluent treatment plant or other means of disposal. Solids should be removed as soon as possible by screening, because prolonged contact of vegetable matter with water results in excessive leaching and increases the problem.

Staff must be provided with adequate and suitable facilities and amenities for changing, washing and eating. There should be lockers for storing outdoor clothing in areas which must be separate from those for storing work clothes. Toilets must not open directly into food-processing areas, all entrances of which must be provided with handwashing facilities arranged in such a way that their use cannot be avoided or unobserved.

The factory layout should be such that processing operations are as direct as possible because the straight-line flow minimises the possibility of contamination of processed or semi-processed product by unprocessed or raw materials and is more efficient in terms of handling. It is also easier to segregate clean and dirty process operations and restrict movement of personnel from dirty to clean areas. While, ideally, the process line should be straight this is rarely possible but there must be no backtracking and, where there are changes in the direction of process flow, there must be adequate physical barriers. It is important that unloading areas for harvested vegetables are sited well away from finished product despatch because of the amount of dirt and debris carried in from the land on tyres of vehicles.

10.5 Equipment

The operational requirements of food-processing equipment have, for many years, been the main concern of equipment manufacturers. In recent years, however, more effort has been put into the hygienic design of vegetable-processing equipment in particular. As for the internal finishes of food-processing areas, there is little guidance concerning food-processing equipment by legislation; the United Kingdom Food Hygiene (General) Regulations 1970 state that articles or equipment with which food comes into contact shall be kept clean, so constructed, be of such materials and kept in such good order, repair and condition as to enable them to be thoroughly cleaned and prevent, so far as is reasonably practical, any matter being absorbed by them and prevent any risk of contamination of the food. There are, in addition, a few National and International Standards relating to specific items such as stainless-steel pipes and fittings for the food industry. In June, 1989, however, a European Communities Directive (89/392/EEC) relating to machinery was adopted; within this is a section concerning essential health and safety requirements for agri-foodstuffs machinery. This states that machinery intended to prepare and process foodstuffs must be so designed and constructed as to avoid any risk of infection, sickness or contagion and then lists a number of hygiene rules relating to materials, design and sanitation. In the United States of America there is a National Standard for Food, Drug and Beverage Equipment (ASME/ANSI F2.1-1986) which is quite detailed and specific. The first section of this standard covers all equipment with respect to materials of construction, sanitary design and construction, as well as safety. Subsequent sections are related to specific products. It is stated under the heading 'Specific Criteria for Fruit and Vegetable Canning and Freezing Equipment' that, except for the initial handling steps, the materials of construction and the sanitary design and construction given in the first section of the standard should be adhered to. Contact materials such as concrete, wood (except for washing equipment) and coated or uncoated carbon steel are permitted for the initial handling steps. Detailed requirements are given for fillers, blanchers, can washers, conveyors, heat exchangers (scraped surface), closing machines and corn huskers with respect to materials of construction and sanitary design and construction. In general, the 300 series stainless steel is the recommended material of construction, that is, the Type 304 or the more corrosion-resistant Type 316 which contains molybdenum.

In an American document entitled 'Sanitary Design' by Milleville and Gelber which was published in *Food Processing*, October 1964 and April 1965, seven basic principles for hygienic design of equipment were outlined. Although these have been quoted by several authors they are still considered to be generally acceptable and are worthy of being repeated:

1. All surfaces in contact with food must be inert to the food under the conditions of use and must not migrate to, or be absorbed by, the food.

2. All surfaces in contact with food must be smooth and non-porous so that tiny particles of food, bacteria, or insect eggs are not caught in microscopic surface crevices and become difficult to dislodge, thus becoming a potential source of contamination.

3. All surfaces in contact with food must be visible for inspection, or the equipment must be readily disassembled for inspection, or it must be demonstrated that routine cleaning procedures eliminate the possibility of contamination from bacteria or insects.

4. All surfaces in contact with food must be readily accessible for manual cleaning, or if clean-in-place techniques are used, it must be demonstrated that the results achieved without disassembly are the equivalent of those obtained with disassembly and manual cleaning.

5. All interior surfaces in contact with food must be so arranged that the equipment is self-emptying or self-draining.

6. Equipment must be so designed as to protect the contents from external contamination.

7. The exterior or non-product contact surfaces should be arranged to prevent harbouring of soils, bacteria or pests in and on the equipment itself, as well as in its contact with other equipment, floors, walls or hanging supports.

Additionally, as suggested by a Campden Food and Drink Research Association (CFDRA) Working Party (9) the following general principles should also be applied:

8. In design, construction, installation and maintenance it is important to avoid dead spaces or other conditions which trap food, prevent effective cleaning and may allow microbial growth to take place.

9. The requirement of guarding machinery to ensure safety in operation may easily conflict with hygiene requirements unless considerable care is taken in design, construction, installation and maintenance.

10. Noise suppression is important in providing acceptable working conditions. However, many noise reducing materials can give rise to microbiological or infestation problems unless care is taken in their selection, installation and maintenance.

11. It is important that the equipment itself is so designed, installed and maintained that it does not cause product contamination. Examples of possible contamination are lagging which may break up or insufficiently secured nuts and bolts. Such hazards should be designed out of the system.

The most obvious indication of improvements in the hygienic design of vegetable processing equipment in recent years has been the increase in the use of mild steel as a material of construction for a wide range of equipment such as de-stoners, blanchers, elevators, size graders, air cleaners and steam peelers. These had previously been constructed from coated or uncoated stainless steel and this change towards the use of stainless steel indicates the increasing awareness of the benefits such as reduced maintenance, durability and ease of cleaning despite the higher initial capital cost. Some examples of equipment now made of stainless steel are shown in Figures 10.3 and 10.4. The

Figure 10.3 An abrasive peeler, for the continuous peeling of root vegetables, made of stainless steel (courtesy Cabinplant (UK) Ltd).

Figure 10.4 A stainless-steel elevator (courtesy Cabinplant (UK) Ltd).

finish of product contact surfaces can affect cleanability because rougher surfaces can provide a better mechanical key for product residues. A maximum-surface roughness value of 1.0 μm Ra is suggested for the surface finish which is a not too exacting requirement when it is known that sheet stainless steel, having a 2D finish, as supplied directly by the steel mills, has a value of about 0.5 μm Ra. Hard plastics are also used as a material of construction but it must be ensured that the material is approved for food contact and is inert to cleaning chemicals. Polypropylene, for example, is not resistant to attack by low levels of sodium hypochlorite, a commonly used disinfectant. Plastics are more easily abraded than stainless steel and this affects their cleanability. Conveyor belts are a common source of contamination, especially if the belting has a carcass of an absorbent material such as canvas or cotton. Polyester is a material now widely used for the carcass of belting because it is stable and non-absorbent. Certain materials must be avoided; these include zinc, lead, cadmium, antimony, plastics containing free phenol, formaldehyde or plasticisers and wood in most circumstances. Copper must be avoided, especially if brine is present.

The internal design is obviously one of the most important factors in the hygienic design of equipment. There should be no internal crevices or 'dead spaces' in which product can be retained or from which product residue cannot be readily removed during cleaning. Internal radii should be as generous as technically possible with a minimum of 5 mm and at least 25 mm for tanks and equipment such as hoppers. There should be no metal-to-metal joints in product contact areas because they are uncleanable and can retain very high numbers of microorganisms. All permanent joints should be continuously welded and the weld ground and polished to a surface roughness value similar to that of the surrounding metal. Where this is not possible a gasket of suitable material should be used but arranged to give a flush, continuous crevice-free surface. Non-toxic solders may also be used to seal joints but it must be ensured that they are not attacked by the cleaning chemicals. Exposed screw threads must be avoided in product areas because they are difficult to clean as are countersunk screws and socket head cap screws. Any fastener in or above product areas must be secure to avoid the possibility of product contamination.

While special attention to the hygienic design of all the equipment in a process line may be unnecessary, although it facilitates cleaning, it is vital to identify those parts of the line which are critical control points. In a canning operation, for example, poorly designed post-process can-handling equipment can result in leaker infection which is the commonest form of spoilage encountered in the canning industry and has been associated with a number of serious food-poisoning incidents. The hygienic design of post-process can-handling equipment is, however, an area which is beyond the scope of this chapter and is well covered in a CFDRA technical manual (10).

Attention should also be paid to the external design of equipment to ensure that it may easily be kept clean. External surfaces should, therefore, be smooth and uncluttered so that spillages, dust and debris cannot accumulate. In particular, electric motors should be fitted with cowls or enclosed such that dust and debris cannot become trapped in the cooling fins. Lids should be fitted to hoppers and vats to protect the contents from foreign body contamination. Hinges, if fitted to lids, must be of a simple design, such as the hook-on type, to facilitate cleaning. Horizontal flat surfaces should be avoided, wherever possible, to aid drainage and horizontal tie bars on frames should be of circular cross-section or, if of square section, turned through 45° to minimise the accumulation of debris. The ends of all tubular or hollow section material must be sealed completely. It must be ensured that painted surfaces which should be avoided above the product area, are not liable to flake and must be suitable to withstand, not only product spillages, but also cleaning chemicals. Gearboxes or geared motors should not be mounted above product areas but if this is unavoidable, then drip trays must be provided to avoid possible product contamination.

It is mandatory that all moving parts of machinery are guarded to satisfy safety requirements but these often conflict with hygiene requirements. Guards must be such that they do not increase cleaning problems, as shown in Figure 10.5, and must be easily removable where necessary for cleaning but fitted with suitable safety devices. All fasteners used to retain removable guards should be captive to avoid the possibility of product contamination.

Figure 10.5 Solid side guards as well as tie-bar at floor level make cleaning difficult.

Permanent open-mesh guarding may be fitted but it must be ensured that it is possible to clean behind it.

10.6 Installation

The installation of the equipment should be given careful consideration if cleaning and maintenance problems are to be avoided. A minimum of 1 m should be allowed all round equipment but preferably at least 2 m. Equipment supported by legs should be at least 200 mm off the floor to allow for cleaning beneath. Heavy equipment may have to be mounted directly on the floor or on a plinth, in which case the base must be completely sealed with a suitable compound. There must also be adequate space left above equipment for cleaning and a minimum of 0.5 m is suggested. Larger clearances around and above equipment may be required for maintenance purposes and this must not be overlooked at the planning and layout stages. As mentioned previously, attention to the supply of services to the equipment must be given in order to avoid cleaning and other problems.

Major problems are often caused by insufficient thought being given to the linking of different items of equipment in a line. This can lead to the 'hold-up' of product caused by unsatisfactory transfers and often compounded by the fitting of deflectors or chutes; organoleptic, taint or microbiological problems may result. An example of an unsatisfactory transfer is shown in Figure 10.6 in which a diced vegetable product is being transferred through right angles from one conveyor to another. A solution is shown in Figure 10.7; the product could be fed on to a conveyor by a vibratory feeder, having an angled end to the tray which would ensure an even distribution over the width of the conveyor belt. Many examples of problems of unsatisfactory transfers, all based on actual situations, are given in a CFDRA technical manual (11), together with suggested solutions.

10.7 Maintenance

Much equipment is not designed for ease of maintenance and if it incorporates components such as seals which will require periodic replacement, then the equipment should be so designed to make this as easy as possible.

Maintenance is often carried out on a breakdown basis and then fitters are called into a process area having had to break off from working in a non-process area. Hence they may be wearing soiled overalls and can unwittingly be the cause of hygiene problems. It is most important that all product is cleared from a machine before a repair is carried out and that it is thoroughly cleaned before start-up. Fitters should be supplied with distinctive containers into which dismantled parts can be placed. While it is appreciated that, due to the pressures of production, some repairs have to be of a temporary nature, they can so often become permanent. Operatives also often make temporary

Figure 10.6 Unsatisfactory product transfer from one conveyor at right angles to another.

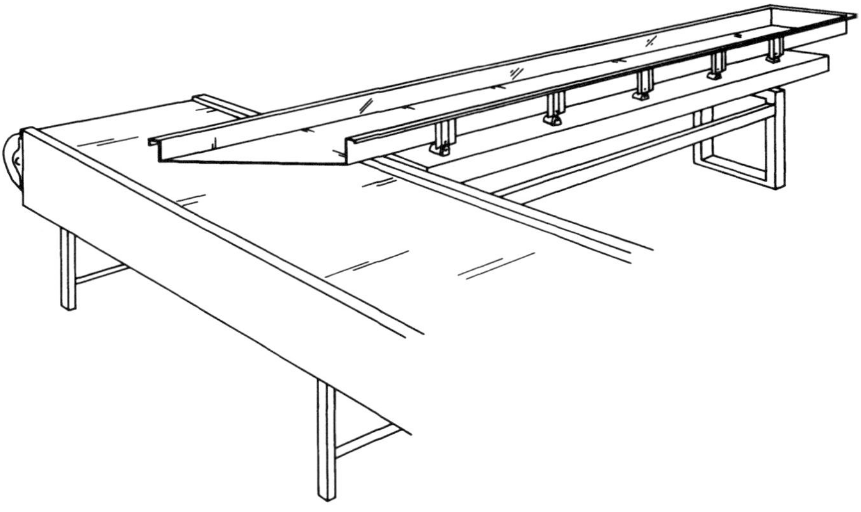

Figure 10.7 A solution to the problem of product transfer through right angles on to a conveyor by the use of a vibratory feeder.

repairs but these must be reported as soon as possible so that a proper repair can be made at the end of production.

There has, of course, to be some degree of planned maintenance of equipment such as greasing of bearings in which case any surplus grease must be removed. Some components such as seals, gaskets and conveyor belts have a limited life and should be replaced periodically otherwise hygiene problems may arise. Worn conveyor belts may contain very high numbers of microorganisms which are impossible to remove during sanitation. Gaskets in pipe couplings harden and crack in time and the first indication of this is a leak which may be temporarily cured by re-tightening the coupling. The gasket should be replaced as soon as possible but it is most probable that all the gaskets in the line have deteriorated, thus it is sensible to replace them all at the same time. Diaphragms in valves have a limited life and these should be replaced periodically in order to avoid costly shut-downs during production. Regular inspections of the plant should be made to check on the condition of painted or galvanised frames as well as the building, because if the repairs are neglected, then damage to the fabric of the structure may result and prove very costly to rectify.

10.8 Cleaning

The regular cleaning of food-processing equipment and premises is an operation vital to maintaining satisfactory product quality. It can only be carried out effectively by trained personnel working to well-documented procedures with the correct equipment and chemicals; the plant must, of course, be propely designed, laid out and maintained. The responsibility for cleaning should be that of a senior manager in the company. Cleaning is not only necessary to meet the hygiene requirements but also for organoleptic and safety reasons as well as to ensure efficient operation of the equipment.

Cleaning can be defined as the removal of any unwanted matter. In vegetable-processing operations most cleaning is wet for both 'open' and 'closed' process equipment; if there are any dry operations, such as the winnowing and sieving of haricot beans or peas, then the cleaning operation should also be dry, preferably by means of vacuum, because the use of compressed air can spread dust and debris over a wide area. Wet cleaning usually involves three forms of energy viz. chemical, heat and mechanical; the chemical energy is provided by the detergent and the mechanical energy by brushes, jet impact or turbulence in the case of in-place cleaning of pipelines. These three energy forms are, to some extent, interchangeable but time is another factor to be considered.

The first stage of any cleaning operation is to remove and dispose of as much gross debris as possible by scraping, brushing or other means. This reduces the amount of detergent required and the load on the effluent plant. There are usually three stages in wet-cleaning procedures – a pre-rinse to remove any residual loose debris, a detergent soak to loosen and hold in suspension soil and/or a bacterial film and then a post-rinse to flush away the loosened soil and detergent. There may, depending on the product and process, be two additional stages – disinfection and another rinse. It is essential, as mentioned previously, that all water used for cleaning is potable; if the water is hard it may need to be passed through a softening plant or softening agents incorporated in the detergent.

In general, 'open' process equipment is not suitable for cleaning in-place due to its design characteristics. Whenever possible, however, consideration should be given at the design stage to the incorporation of spray jets or other cleaning devices so that the cleaning operation can be automated, hence removing the human element and reducing labour costs. Such arrangements can be applied to equipment such as conveyors and bucket elevators where they often operate continuously. Much of the equipment has to be cleaned manually, either by scrubbing or the use of lances delivering water with or without detergent. Some equipment has to be dismantled for cleaning due to its complexity.

In the past, large-bore hoses connected to the mains supply have been widely used to supply a high volume of water at a relatively low pressure. This has often been the main means of cleaning, especially for the equipment used in the preliminary stages of vegetable-processing operations. It is particularly wasteful as it was a common sight to see open-ended hoses left running for long periods. A high volume of low-pressure water is an effective method for moving large quantities of debris, if unavoidable and for pre- and post-rinsing of large areas. It is most important, therefore, to fit self-closing valves to hose outlets to minimise wastage. Where a high volume of low-pressure hot water is required stream/water mixers are often used; these units may also incorporate devices for detergent injection.

High-pressure/low-volume systems are now much more widely used because much less water is required due to the increased mechanical energy of the high-velocity jet. Detergents and disinfectants may be dosed locally or centrally. A high-pressure jet, having a flowrate of 0.7 litre/min, can give the same impact/unit area as a low-pressure hose at 3.7 litre/min; the cost savings in steam, water and effluent of high-pressure/low-volume compared with low-pressure/high-volume are of the order of 85%. High-pressure/low-volume systems operate at pressures up to 130 bar; while the mechanical energy is very high, water at this pressure can be both dangerous and harmful to equipment by washing grease out of bearings, damaging electrical equipment and removing grouting from tiled floors. Pressures that are excessively and unnecessarily high can also cause problems by scattering soil over a wide area, possibly onto already cleaned surfaces, as well as the production of aerosols which can cause lung irritation to the operatives if detergents are contained (Figure 10.8). It is advisible to conduct cleaning trials using a variable pressure cleaning machine before deciding on the pressure rating of a system; this should be as low as possible for effective cleaning and in many instances pressures of the order of 30 to 40 bar are quite adequate. The high-pressure water may be supplied by mobile electrical pump units, wall-mounted units or a central pumping and distribution system having fixed take-off points. A mobile unit, such as that shown in Figure 10.9, has the capacity to supply a

Figure 10.8 High-pressure jet cleaning of equipment.

single lance; it has to be connected to water and electrical supplies; a detergent and possibly a disinfectant in containers of about 5 litre capacity can be carried in a rack attached to the machine. Wall mounted units, as shown in Figure 10.10, again supplying a single lance, are permanently connected to the electrical and water supply; the detergent containers can also be incorporated in the frame as can a hose reel. Both of these units can be supplied with water at temperatures up to about 80 °C; the outlet pressure is infinitely variable up to 100 bar and the flowrate up to 12 litre/min. Central cleaning systems, having a permanent distribution system with fixed take-off points sited throughout the factory, are supplied with high-pressure water from a pumping unit, usually located outside the processing area. There can be more take-off points than the pumping unit is capable of supplying but the number of lances that can be used at any one time is limited by the maximum capacity. Cleaning chemicals may be distributed from a central remotely sited location to each take-off point, or supplied by containers supported by a wall rack located adjacent to each take-off. This rack can also accommodate a lance and hose; alternatively, lances and hoses may be moved and plugged into different take-off points. A single large pump may be used to supply a central system but in the event of failure the factory may be left without any means of cleaning. It is recommended that there should be at least two pumps because half the capacity is better than none at all. One or two large pumps may not be cost effective especially as the demand on the supply from a central system can be variable. Multipump systems, such as shown in Figure 10.11, have been developed to supply only the quantity of water required. The unit shown incorporated six pumps and a

Figure 10.9 A mobile high pressure cleaner (courtesy K.E.W. Industry Ltd).

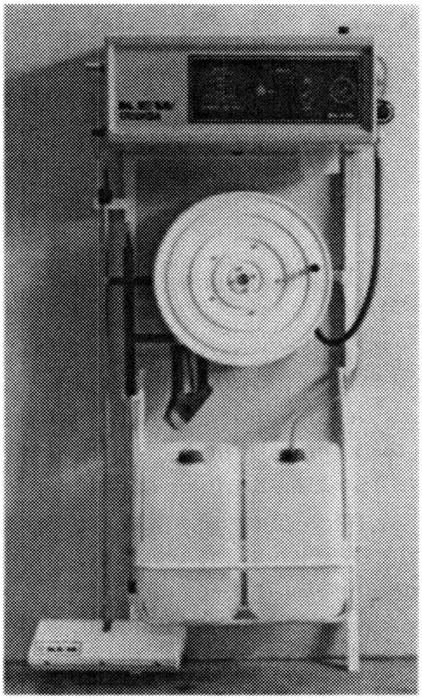

Figure 10.10 A wall-mounted high-pressure cleaner and stand for accessories (courtesy K.E.W. Industry Ltd).

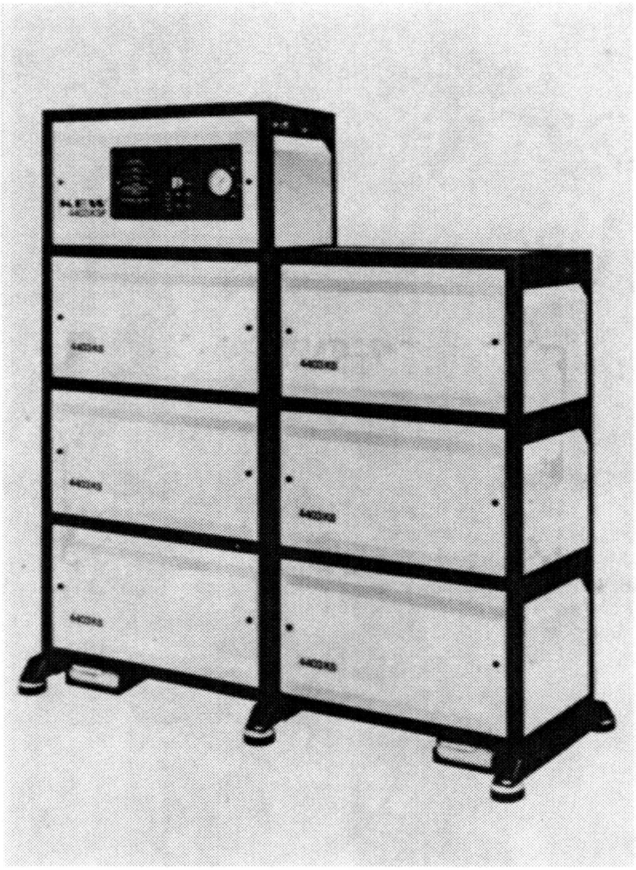

Figure 10.11 A modular high-pressure pump unit to supply a cleaning distribution system (courtesy K.E.W. Industry Ltd).

controller; it can supply water at pressures of up to 160 bar, if required, to a maximum of six lances and is controlled automatically by demand so that additional pumps are brought into operation as more lances are used, up to the maximum number. Hence, if during a shift it is necessary to use only one lance occasionally, then only one pump will be started, thus saving energy. Single lances are usually used when chemicals are not required. Double lances incorporate two nozzles, one for dispensing chemicals at low pressure and the other water at high pressure. Lances for applying foam have been developed but a different injector is required from that for dosing of liquid chemicals. The application of detergent as a foam is particularly suitable for cleaning large areas and vertical surfaces such as walls and ceilings, as well as the exterior surfaces of process equipment and product contact surfaces which are not too heavily soiled. The advantage of foam is the extended contact time of up to about 10 min. The method does lack one of the three energy forms involved in the cleaning operation – that of heat, but it is claimed that chemical utilisation is far superior to conventional methods. Quite large quantities of water may, however, have to be used for rinsing. Detergent may also be applied as a gel and this may remain in contact with soiled surfaces for much longer periods of up to several hours, thus allowing for penetration and loosening of very tenacious soils.

Some process equipment has to be dismantled for cleaning due to its complexity. The cleaning of the components should be carried out in a tank containing hot detergent. After soaking, the loosened soil can be removed by brushing or spraying. Very small items should be contained in a wire basket. After adequate rinsing, parts may be placed in another tank containing a disinfectant, if required, but would have to be rinsed again. Cleaned components should then be placed on racks to dry prior to re-assembly. It is most important not to place dismantled parts of equipment on the floor at any stage or clean them over drains as this may not only result in increased microbiological contamination but in mechanical damage.

'Closed' process equipment consisting of pipework, valves and vessels has to be cleaned in-place (CIP) as any other means would be impractical. CIP is the technique of cleaning by circulating water and chemicals through the equipment without the need for dismantling it. It is, however, vital that the process equipment selected is designed for in-place cleaning and that it is properly installed with appropriate falls to drain. The equipment must also be of good hygienic design and constructed of materials capable of withstanding the higher temperatures and stronger chemicals that can be used for CIP by comparison with manual cleaning. The selection of a CIP unit will depend upon many factors such as the size of the plant to be cleaned and the soil loading. The CIP pump should be sized so as to give a mean velocity of flow in pipelines of about 1.5 m/s (12); this is particularly important if the soiling is tenacious otherwise cleaning times may be lengthy. The three main types of CIP unit are the single use, re-use and multi-use. The single use unit is the simplest and is suitable for small plants or for cleaning heavily soiled equipment because the cleaning solution is used only once. Re-use units are more complex and quite large as they may comprise up to six tanks because cleaning chemicals and water are recovered and held for re-use. Multi-use units combine the best features of both systems and require only two tanks, one for detergent and one for water recovery. The system is widely used for in-place cleaning applications due to its versatility of being able to provide a wide range of thermal and sequencing combinations; Tamplin (13) has covered the subject in considerable depth.

In many areas of the factory the floor will require to be cleaned daily and although water alone is frequently used it will not prevent the gradual build-up of soil. One of the three energy forms required for cleaning, that of heat, is unavailable because on contact it is almost completely absorbed by the floor material itself. Floor cleaning is dependent, therefore, on the other two energy forms – chemical and mechanical, of which the latter may be produced by pressure jets or rotary scrubbing machines. Low-pressure/high-volume hoses are useful only for pre- and post-rinsing. High-pressure/low-volume lances may be used for floor cleaning but soil and possibly microorganisms may be scattered over a wide area; such problems can be avoided by the use of floor cleaning attachments. These incorporate a horizontal spray bar containing several downward facing nozzles enclosed by a hood fitted with wheels. The use of high-pressure jets to clean drainage channels should be avoided because of the strong possibility of contaminating the processing area and the equipment. Walls probably require cleaning less frequently than floors; foam, because of its extended contact time and high-pressure jetting is often used. There are other wall-cleaning methods such as the use of a rotating brush supplied with detergent and fitted to the end of an extension rod up to about 5 m long. In another system the cleaning fluid is delivered to a special sponge head, fitted to the end of a telescopic handle also having a reach of about 5 m and then vacuumed back into a dirty water tank, thus avoiding drips on the floor. Any internal framework of the roof structure should be kept free from dust and if it is necessary to wet clean the ceiling then foam and high-pressure jetting are often used.

References

1. Cattell, D., *Specialist Floor Finishes: Design and Installation*, Blackie, Glasgow and London, 1988, p. 49.
2. Hayes, P.R., *Food Microbiology and Hygiene*, Elsevier Applied Science Publishers, London and New York, 1985, p. 175.
3. Katsuyama, A.M. and Strachan, J.P. (eds.), *Principles of Food Processing Sanitation*, The Food Processors Institute, Washington, D.C., 1980, p. 103.
4. Imholte, T.J., *Engineering for Food Safety and Sanitation*, Technical Institute of Food Safety, Crystal, Minnesota, 1984, p. 92.
5. Graham-Rack, B. and Binsted, R., *Hygiene in Food Manufacturing and Handling*, Food Trade Press Ltd, London, 1973, p. 65.
6. Holdsworth, S.D., *The Preservation of Fruit and Vegetable Food Products*, Macmillan, London, 1983, p. 146.
7. Packman, P.E., 'Hygienic factory installations' in *Design for Hygiene – The Changing Scene in the Food Industry*. The Institution of Mechanical Engineers, London, 1987.
8. Hulland, E.C., 'Hygienic handling and the influence of raw material condition', in *Hygienic Design and Operation of Food Plant* (ed. R. Jowitt), Ellis-Horwood, Chichester, 1980, Chap. 9.
9. *The Principles of Design for Hygienic Food Processing Machinery*, The Campden Food and Drink Research Association, Technical Memorandum 289, 1982.
10. *Hygienic Design of Post Process Can Handling Equipment*, The Campden Food and Drink Research Association, Technical Manual 8, 1985.
11. *Hygienic Design of Food Processing Equipment*, The Campden Food and Drink Research Association, Technical Manual 7, 1983.
12. Timperley, D.A., 'The effect of Reynolds number and mean velocity of flow on the cleaning in-place of pipelines', in *Fouling and Cleaning in Food Processing*. (eds B. Hallström, D.B. Lund and Ch. Trägårdh), Lund Univ., Sweden, 1981.
13. Tamplin, T.C., 'CIP technology, detergents and sanitizers', in *Hygienic Design and Operation of Food Plant* (ed. R. Jowitt), Ellis-Horwood, Chichester, 1980, Chap. 11.

Index

accelerated freeze drying (AFD) 159
acceptable quality level (AQL) 250
acetic acid 6, 172, 176
acidity groups 70–71
Aeromonas hydrophilia 147
agrochemicals 28
air drying 156
Alaska peas 35
alcohol insoluble solids (AIS) 18–19
aluminium
  cans 191, 218
  foil trays 217
amino acids 1–3
ammonia 109
anticholinesterases 4
Appert, Nicholas 197
Arrhenius rate equation 103, 146
ascorbic acid 130
aseptic packaging 84–97, 218–224
  bulk systems 93, 223
  cans and bottles 91
  cartons 93
  cups 91
  filling systems 92–95
  sachets and pouches 91
asparagus 51
assessors
  sensory 247
  autolysis 5
Automatic Defect Removal (ADR) 49
a_W 128, 149, 155

Bacillus cereus 148
bag-in-box 96
beading, of cans 190, 196
beans, broad 17
beans, green 37–38
  freezing 118–119
  maturity 19–21
  quality 233
  trimming 51
beetroot, peeling 54
benzoic acid 6, 180
biological oxygen demand (BOD) 43–44, 53, 260
blanching 6, 59–67, 77–78, 107–108, 146, 161, 227, 235–236
  effect on pesticide residues 236
  hot-gas 65
  in-can 65
  microwave 65
  pipe 61
  steam 62
  vacuum steam 64
  water 61
boron 24
botulinum cook 71–72
botulism 148
brassicas 38–39
  drying 162
brining 163–170
Brix 244
broccoli
  chill storage 127
  freezing 119
brown onions 167–168
bruising of vegetables 234
Brussels sprouts 16, 18, 22, 39
  freezing 120
BS 5750 186–187, 204

cabbage 38
calcium 23
calcium chloride 130
Campden Food and Drink Research Association (CFDRA) 15, 209, 251, 262, 264
Campylobacter fetus 148
Candida 183
canning 6, 76–84
cans 74–75, 190
  corrosion 197
  manufacture 192

carbon dioxide 104, 133, 229
carrots 22, 36–37
  drying 162
  freezing 119
  root shape 37
  specification 248
catalase 60, 107
cauliflower 22, 39
  blanching 66
  brining 165–166
  freezing 119
ceiling 257
celery 22
chilled vegetables
  distribution 144
  packaging 228
  processes 138
chilling 6, 123–153
  injury (CI) 127
  methods 139–140
  speed of 131–132
cleaning 268–274
  dry 45–46
clean-in-place (CIP) 273
clean room principles 140–141
chlorine 129, 130, 209, 235
chlorofluorocarbons 104, 109
citric acid 130
Clostridium botulinum 7, 69–71, 148
Clostridium perfringens 148
closures, for glass 202
  non-venting 205
  venting 204
Codex Alimentarius Commission 29, 242, 248, 252
colours 181, 242–243
Combibloc 221
combination treatments 126, 180
composite films 129
compressors 108
condenser 108
conduction packs 72
consumers 232, 240
contaminants 184
contracts, for raw materials 14
Controlled atmosphere packaging (CAP) 133–134
convection packs 72
conveyors 267
  belt 44
  vibratory 44–45
cook-chill 142–143
cooking 6
copper 24
crop
  husbandry 21–30
  maturity 17–19
cross contamination 132
cryogenic fluids 110
cucumbers, pickled 166–167
cullet 200
cultivar 16
cyanogens 4

debrining 170
Decimal Reduction Time (*D* value) 69
defects 15, 49, 234, 248
dehydrated vegetables, packaging 277–278
dehydration 6, 13, 154–163, 227–228
dehydrofreezing 160
Department of Health and Social Security 131, 142
dicing 57
diseases 24–30
  resistance to 27
doors 257
double seams, in cans 194
drainage channels 255–256
draw redraw cans (DRD) 192, 195
draw wall-ironed cans (DWI) 192, 196, 218

easy-open ends 197
effluent 260
electronic nose 244
energy
  electrical 42
  mechanical 42
enzymes 74, 130–131, 181
  inactivation 60
  inhibitors 5
equipment, hygienic design 260
Escherichia coli 147
essential amino acids 2
ethylene vinyl alcohol copolymer (EVOH) 212, 215
European Community 252
evaporator 108
exhausting 78
expansion valve 108
extraneous vegetable matter (EVM) 25, 255

factory 255
FAO/IAEA/WHO Expert Committee 134
fatty acids 2–3
Federal Agency 251
fermentation 6
  brining 164–165
filling 78
flame sterilisation 83
flexible pouches 75–76
floors
  ceramic tiles 255
  concrete 255
flumes 44
Food Hygiene Regulations (UK) 255

forehearth and feeder, glass 200
foreign matter 15, 184
freeze drying 159, 162
freezerburn 224
freezers
  batch-plate 112
  blast-air 111
  continuous 115
  continuous tunnel 115
  fluidised bed 115
  semi-continuous 114
  spiral belt 115
freezing 6, 13, 101–122
  mechanics of 108–111
freon 104
freshening 170
fresh vegetables, packaging 228
frozen vegetables, packaging 224
furnace, glass 200

gherkins, pickled 166–167
glass containers 75, 186, 197–208
  handling 206
  manufacture 198
  properties 198
  surface treatment 202
glucosinolates 4, 240
good manufacturing practices (GMP) 149–150
grading 77
grazing 231
greenhouse effect 145

harvesting 30–32
hazard analysis critical control point (HACCP) 150, 180
heat
  exchange 85, 104
  penetration 72
  treatment 5, 130–131, 138
herbicides 28
hexachlorocyclohexane (HCH) 30
high-pressure/low-volume cleaning systems 269
high-temperature short-time processing (HTST) 74, 84, 100
humectants 128
hurdle treatments 126, 180
hydrocooling 140
hydrogen peroxide 130
hydrometer 163
hygiene 132
hygienic design, principles 261–262
hyo-on 127

IAEA/FAO/WHO Expert Committee 134
immersion freezers 110
immunoassay 246
Independent Section (IS) 201
individual quick blanching (IQB) 63, 78
individual quick freezing (IQF) 117
inspection 48
installation of equipment 266
Institute of Food Technology 144
intermediate moisture food 160
International Institute of Refrigeration (IIR) 242
irradiation 5–7, 134–137, 180
ISO 9000–9004 187
isotonic solution 62

jojoba oil 129
Jupiter system 86

lacquers 191
lactic acid 164, 166, 168, 176
Lactobacilli 183
lathyrogens 4
legumes 18
  for drying 161–162
lehr 202
Leuconostoc brevis 169
Leuconostoc mesenteroides 169
Leuconostoc plantarum 169
lifestyles 231
lighting 257
linoleic acid 2
linolenic acid 2, 244
lipid coatings 129
lipoxygenase 60, 108
Listeria monocytogenes 7, 147
location, of factory 254
low-pressure, high-volume cleaning systems 269

macronutrients 1
Maillard reaction 179
maintenance, of equipment 266
malathion 30
manganese 24
maturity 16–17
maturity indices 18–21
Maximum Residue Limits (MRL) 28–29
micronutrients 1
microwave sterilisation 83
minimal processing 125
modified atmosphere packaging (MAP) 133–134, 149
Moniliella acetoabutans 180, 183

near infrared spectroscopy (NIRS) 245
nitrogen 24, 104, 229
nucleation 105
nucleic acids 5

nutrients
  loss on blanching 66–67
  retention in thermal processing 97–98
nutrition, of plants 23–24
nutritional value 1–3, 8, 146–147, 240–241

off-flavours 30, 234
Ohmic heating 88
olives 169
onions 53
  pickled 167–168
osmosis 6, 106
osmotic drying 160
oxalates 4
oxidation 238
oxygen 133, 229
  permeability of films 135

packaging 7–8, 132
  economic benefits 188
  long-term benefits 189
  marketing benefits 188
  requirements 187–188
parison 202
pasteurisation 131, 137–139, 207
  of acid products 176–178
pathogenic bacteria 147–150
peas 18, 34–36, 47, 254
  cultivars 36
  freezing 118
  harvesting 30
  quality 233
  size-grading 36
peeling 51, 77
  hot water 54
  lye 53–54
  mechanical 52
  steam 55, 235
permeable packaging, films 133
peroxidase 60, 107
pesticides 27–30
pests 24–30
PET 206, 212, 215, 217
pH 6, 70–71
  modification 127
  of vegetables 71
phenolics 4
pheophytin 236
pickling 171–176
pipes 259
plant nutrition 23
plant population 22
plastics
  bottles 215
  containers 212
poyethylene terephthalate *see* PET
polypropylene 212, 264
polysaccharide coatings 129
polyvinylidene chloride (PVDC) 212
potatoes 18, 22, 39–41
  cutting 59
  drying 162
  freezing 120
  inspection 49
  processing requirements 39–40
  skin removal 56
  storage 33
practical canning stage (PCS) of peas 18–19
preparation procedures 42–68
  for dried vegetables 160–161
preservation 5
  effects 8
  processes 6
Preservation Index (PI) 172–173, 178
preservatives 178–180
protein coatings 129
proteins 1, 4
psychrotrophic microorganisms 127
pumps 44

quality 230–253
  definitions 230–232
  effects of processing 234
  factors affecting 16–34
  measurement by instrument 241–246
  raw material 15
  supply of 14
  visual 15

raw material quality 15–34, 233–241
recycling, of cans 197
red cabbage 168
residues 28–29
retort pouch 208–212
  post process handling 209
retorts
  batch 79–81
  continuous 81–83
risk zones 132
rod washer 47
rootcrops 47
  drying of 162

Saccharomyces 183
safety, in chilled foods 147–150
Salmonella 147
salt 163–170
  stratification 170
sanitary
  packaging 140
  processing 132
sauerkraut 168–169
  fermentation 169

sensory evaluation 246–248
shelf-life 147
  factors affecting 184
silverskin onions 167–168
site, of factory 254
skin removal 56
slicing 57
sodium metabisulphite 6
sodium nitrate 6
sodium sulphite 6
soil type 22
sorbic acid 130, 180
sous-vide 142–143, 147
specifications 14–15, 248–252
spectrophotometer 242
spinach, freezing 120
spoilage, of acid products 182–184
spray drying 157
stainless steel 263–264
Staphylococcus aureus 148
sterilisation, of glass containers 207
Steripart technique 89–90
storage 32–34
  canned products 98–100
  chilled foods 126–127
  frozen products 121–122
  potatoes 33
sulphur dioxide 129, 179
superchill 127
surface coating systems 128–129
surface dipping systems 129–130
sweet corn 53
  freezing 120

taints 30
Tenderometer 19–20, 161, 245
Tetra-Pak 219–220
texture measurement 245
thermal death-rate 70
time–temperature indicators 145
time–temperature tolerance (TTT) 239
tin-free steel (TFS) 191, 203
tinplate 190
tomatoes 48
  inspection 50
  paste or purée 173
  peeling 54
toxins 4
trays, aluminium 76
trimming 51
tubes, collapsible 218

ultra high temperature process (UHT) 84
United States Department of Agriculture (USDA) 251–252

vacuum cooling 140
vacuum drying 158
vegetable processing 5
  freezing 105–107
  future 10
  processes 6
  reasons for 3
ventilation 258
Vibrio parahaemolyticus 147
vinegars 175
viners 30–31
vitamins 1–2
vitamin C 2, 8–10, 98, 146, 239, 241
  loss on blanching 66–67
vitamin E 2–3, 10

walls, of factory 256–257
walnuts 168
washing 46–48, 77, 235
water activity *see* a_W
water supply 259
wax coatings 129
weed control 24–30
welded cans 13, 192
WHO/FAO/IAEA Expert Committee 134
windows 257

Yersinia enterocolitica 147

z-values 72–74

Lightning Source UK Ltd.
Milton Keynes UK
UKOW03n0613070813

214967UK00001B/92/A

9 780471 198598